Partnerships for a Common Purpose: Cooperative Fisheries Research and Management

Funding for this book was provided by

NOAA National Sea Grant Office
NOAA National Marine Fisheries Service
American Fisheries Society
NOAA National Ocean Service
Northeast Consortium
Columbia River Inter-Tribal Fish Commission
North Pacific Research Board
Manomet Center for Conservation Services
Sea Grant Fisheries Theme Team
USGS Biological Resources Discipline
Gulf of Maine Research Institute
Bristol Bay Economic Development Corporation
Bristol Bay Science and Research Institute
Great Lakes Fishery Commission
Continuing Studies in Science, Simon Fraser University
Alaska Longline Fishermen's Association
AFS Marine Fisheries Section
Maine Department of Marine Resources
Alaska Marine Conservation Council
Northwest Power and Conservation Council
Marine Conservation Alliance
Bristol Bay Native Association
Centre for Community Based Management, St. Xavier University
Oak Ridge National Laboratory

Partnerships for a Common Purpose: Cooperative Fisheries Research and Management

Edited by

Alesia N. Read
American Fisheries Society, 5410 Grosvenor Lane, Suite 110
Bethesda, Maryland, 20814, USA

and

Troy W. Hartley
Northeast Consortium
Resource Economics and Development, University of New Hampshire
310 James Hall, 56 College Road, Durham, New Hampshire, 03824, USA

Symposium 52

Proceedings of the American Fisheries Society/Sea Grant Symposium
Partnerships for a Common Purpose:
Cooperative Fisheries Research and Management
Held in Anchorage, Alaska, USA
13–14 September 2005

American Fisheries Society
Bethesda, Maryland
2006

Suggested citation formats are

Entire book

Read, A. N., and T. W. Hartley, editors. 2006. Partnerships for a common purpose: cooperative fisheries research and management. American Fisheries Society, Symposium 52, Bethesda, Maryland.

Chapter in book

Mattes, W., and N. Kmiecik. 2006. A discussion of cooperative management arrangements within the Ojibwa ceded territories. Pages 163–168 *in* A. N. Read and T. W. Hartley, editors. Partnerships for a common purpose: cooperative fisheries research and management. American Fisheries Society, Symposium 52, Bethesda, Maryland.

Photographs on front cover: photos numbered 1–17, clockwise starting at the top left corner: numbers 1, 4, 10, and 12, photographed by Wolcott Henry, gathered from the SeaWeb Marine Photobank; numbers 5, 8, 13, and 16, National Oceanic and Atmospheric Administration, Department of Commerce photo gallery; numbers 6, 7, 11, and 14, Ken La Valley, New Hampshire Sea Grant; and numbers 2, 3, 9, 15, and 17, Northeast Consortium photo library.

Printed in the United States of America on acid-free paper.

Library of Congress Control Number 2006931346

ISBN 1-888569-85-9
ISSN 0892-2284

American Fisheries Society Web site: *www.fisheries.org*

American Fisheries Society
5410 Grosvenor Lane, Suite 110
Bethesda, Maryland 20814-2199
USA

Contents

Preface

In recent years, there has been a growing interest in greater stakeholder involvement in the fisheries research and management process. Direct multiple stakeholder involvement in these processes is generally referred to as cooperative research and cooperative management. The concepts of cooperative research and management are not new; there are many examples of successful implementation of cooperative programs and cases where working cooperatively in sustainable natural resource management can be extremely effective. The symposium explored and evaluated what difference cooperative research and management make, how successful cooperative programs work, and what future opportunities exist, in terms of scientific, social, cultural and economic impacts.

The American Fisheries Society (AFS)/Sea Grant-sponsored symposium, entitled "Partnerships for a Common Purpose: Cooperative Fisheries Research and Management," took place during the 135th Annual AFS meeting in September 2005 in Anchorage, Alaska. This is the fourth in a biennial series of special symposia held in conjunction with the annual AFS meeting. The series has grown to include many sponsors over the years and has become a landmark component of AFS meetings. Symposium topics are meant to be broad and interdisciplinary in nature, with the goal of participants leaving with a set of action items to direct future research. The series has been a great success over the years with "Fish Habitat: Essential Fish Habitat and Rehabilitation" (1998, Hartford, Connecticut), "Fisheries in a Changing Climate" (2001, Phoenix, Arizona), and "Aquatic Protected Areas as Fisheries Management Tools" (2003, Quebec, Canada) all published as special symposia proceedings.

Continuing in the vein of addressing relevant, timely, and critical topics in the field of fisheries, we designed a symposium focused on cooperative fisheries research and management. It was difficult to fit as much information as we wanted into a two-day symposium, which was why we focused on practical advice and lessons learned, on how to build successful working partnerships rather than detailing empirical research and recent project results. In order to highlight the diversity of cooperative programs that are underway, we invited posters to be presented at an accompanying social. This allowed for networking and discussion among participants beyond the panel discussions. We were well aware that we could have spent more than a week on either cooperative research or cooperative management alone. In order to cover as much high quality information as possible, the steering committee made sure to choose speakers that would ensure geographic equity and include representatives from many stakeholder groups in the program. The symposium incorporated a diverse set of panel discussions, each panel consisting of an industry representative, a manager, a scientis, and a community participant. This format offered a broad range of perspectives, with panelists and keynote speakers selected for their many years of expertise and first-hand experience in this field. The two-day symposium was filled with rich discussion and presentation topics. The discussions pushed forward our understanding of the critical issues facing cooperative fisheries research and management today.

There is still much more work to be done before cooperative research and cooperative management are accepted as a default paradigm in fisheries. The benefits realized by successful cooperative programs are undeniable in terms of leveraging resources, maximizing while not duplicating efforts, all while accomplishing the common goal of maintaining sustainable fisheries resources. Whether your job is to catch fish for revenue, to estimate stock population size, or to set the regulations and quotas on how many fish can be caught, you will get more bang for your buck by opening lines of communication and working together.

I am indebted to the keynote speakers and panelists who took time out of their busy schedules and were eager to participate and share their vast knowledge and experiences both at the symposium and within this publication. The Anchorage meeting organizers made things run smoothly and were incredibly helpful in accommodating all of our needs. I thank Bill Wilson, Eric Knudsen, and Christine Fletcher for all of their hard work at this record-breaking, large meeting!

Many organizations provided invaluable support, both monetary and in-kind, which made the symposium possible: AFS Marine Fisheries Section; Alaska Longline Fishermen's Association; Alaska Marine Conservation Council; American Fisheries Society; Bristol Bay Native Association; Bristol Bay Science and Research Institute; Centre for Community Based Management; St. Francis Xavier University; Columbia River Inter-Tribal Fish Commission; Continuing Studies in Science, Simon Fraser University; Great Lakes Fishery Commission; Gulf of Maine Research Institute; Maine Department of Marine Resources; Manomet Center for Conservation Sciences; Marine Conservation Alliance; NOAA National Marine Fisheries Service; NOAA National Ocean Service; NOAA National Sea Grant Office; North Pacific Research Board; Northeast Consortium; Northwest Power and Conservation Council; Oak Ridge National Laboratory; Sea Grant Fisheries Theme Team; and USGS Biological Resources Discipline.

I would like to extend my sincerest thanks to the members of the steering committee who were instrumental in developing the program and content: Emory Anderson, Dorothy Childers, John Chythlook, Chuck Coutant, Paula Cullenberg, Lisa Desfosse, Mark Chandler, Carlos Fetterolf, Patricia Gallaugher, Deirdre Gilbert, Chris Glass, Gary Graham, Troy Hartley, John Hoey, John Kearney, Linda Behnken, Gary Matlock, Clarence Pautzke, Robert Pomeroy, Gus Rassam, Phillip Roger, Peter Rubec, Laura Taylor Singer, Bob Stickney, and Steve Waste. Also, we are particularly appreciative to Ron Baird, director of the National Sea Grant Office, for his continual support and encouragement of the fellowship program. Special appreciation for the moderators who helped in making everything flow smoothly: Ralph Rayburn, Troy Hartley, Rollie Barnaby, Paula Cullenberg, John Schwartz, and Jim Murray.

Troy and I would also like to thank Olivia Saunders and Maureen Norton for help they provided in formatting. Also, Aaron Lerner for technical guidance and review of the proceedings. Special thanks also to Debby Lehman, the AFS book coordinator.

Even though I thanked them above for their service on the steering committee, Emory Anderson, Gus Rassam, and my coeditor Troy Hartley provided substantial guidance and assistance along the way. Troy certainly went above and beyond what was asked of him, and his passion for the subject matter was truly infectious. Emory continuously imparted a stable sounding board that allowed me to be creative while maintaining focus. Gus has been an incredible influence as a mentor and an employer. I sincerely enjoyed my fellowship, and I know this is substantially due to Gus's management style. I am so thankful to have had this opportunity to complete this fellowship and to have met and worked with as many people as I did. From the stakeholder interviews to compiling this publication, this has truly been the most rewarding professional experience I have had thus far in my life. I have learned volumes and continue to do so as I continue to work for AFS. I am grateful to the staff of AFS as well; they have provided a comfortable and enriching work environment that embraced me from my first day. I hope the readers of this publication learn and gain as much knowledge and enjoyment as I did from putting it together.

ALESIA N. READ
Coeditor

AFS Perspective: Introduction for Symposium Proceedings on Cooperative Fisheries Research and Management

Barbara A. Knuth
American Fisheries Society Past-President 2004–2005
DNR 117 Fernow Hall, Cornell University
Ithaca, New York 14853, USA

Achieving the mission of the American Fisheries Society, "to improve the conservation and sustainability of fishery resources and aquatic ecosystems by advancing fisheries and aquatic science and promoting the development of fisheries professionals," can best be achieved through partnerships. I think most would agree with this claim; improving conservation and sustainability of fisheries resources is a daunting task, one that grows more complicated each year as the number of resource users (such as fish consumers) grows, the quality of fish habitat plummets, and the numbers of many fish stocks decline. Many of the factors influencing the sustainability of fisheries resources are out of the hands of fishery managers.

Partnerships are not an unfamiliar concept to fisheries professionals. Members of the American Fisheries Society (AFS), by the very nature of their professions, encourage partnerships between AFS and their employing organizations, whether in the private for-profit, not-for-profit, or government sectors. Units of AFS, particularly chapters, partner at various levels with local community, regional, or state-level citizens organizations and other stakeholder groups to develop and implement educational programs, on-the-ground activities such as habitat improvement projects, and a wide variety of other outreach efforts. It should be no surprise then that the larger fisheries research and management enterprise, however broadly defined, should also benefit through cooperative efforts, involving fisheries stakeholders, managers, and scientists.

The theme for the 135th AFS annual meeting in Anchorage, Alaska was "Creating a Fisheries Mosaic: Connections Across Jurisdictions, Disciplines, and Cultures." This theme was intended to highlight the importance of partnerships, connections, and collaborations necessary to address the complex fisheries resource management challenges we face now and into the future. The joint AFS/Sea Grant-sponsored symposium, "Partnerships for a Common Purpose: Cooperative Fisheries Research and Management," focused this theme even further. The symposium organizers and participants have provided a valuable service to the fisheries profession by asking tough questions about what "cooperative" really means when used in reference to "management" and "research" and by providing insights, answers, and even further questions about the implementation of cooperative research and management, its potential for success, the barriers it faces, and the opportunities we have into the future to use the principles of cooperative research and management to improve the way in which we advance the conservation and sustainability of fishery resources and aquatic ecosystems. Let us learn from the insights shared and develop approaches that help leverage the commitment and perspectives from a variety of fisheries stakeholders as we tackle conservation challenges into the next century.

Foreword

RONALD C. BAIRD[1]
National Sea Grant College Program
National Oceanic and Atmospheric Administration
1315 East-West Highway, SSMC-3, Room 11716
Silver Spring, Maryland 20910, USA

This is the fourth volume of the proceedings of a biennial special seminar series initially sponsored by the National Oceanic and Atmospheric Administration's (NOAA) Sea Grant Program, NOAA Fisheries, and the American Fisheries Society. Now, along with these partners, there are a host of new sponsors, a fitting testimony to the success of the symposia. The founders had a vision for this series—something different than a strictly science-focused symposium. The intent was to address important issues in contemporary fisheries science that have significant management implications. Moreover, the idea was to introduce the management context by incorporating the views of management professionals and various constituencies and stakeholders.

It is this forum that provides the framework for scientific presentations in the proceedings and informs emerging public policy. The other essential aspect of this vision is the rapid synthesis and publication of results so essential to policy.

Our first symposium addressed the issue of essential fish habitat as defined in the Sustainable Fisheries Act (Magnuson/Stevens), a topic that remains timely today. Our second dealt with climate and fisheries, an issue at the forefront of environmental concerns. The third considered aquatic protected areas, an issue of global concern today as we seek remedies for distressed fish stocks. We continue that tradition in this volume by considering another emerging issue in fisheries management—cooperative fisheries research—a partnership of fishers and scientists focused on developing solutions to current management problems.

It is my pleasure to acknowledge four individuals for their considerable contributions to this series since its inception: Paul Brouha, Rollie Schmitten, Carlos Fetterolf, and Emory Anderson. We owe them a debt of gratitude for making this series possible. Finally, a word of thanks to Alesia Read, one of the editors and a Sea Grant fellow, whose hard work is evident here.

While the concept of sustainability provides the conceptual framework driving current fishery management practice, we are a long way from achieving sustainable fisheries globally. Increasing perturbation of the environment by human activities clearly calls for better performance by our institutions of governance and new methods for managing fisheries. In the case of cooperative research, we have an increasingly important and often contentious issue from both a management perspective and the fisheries science basis upon which the concept rests. As with any management approach, to be effective not only requires good, sound science as a basis for decision making, but also an appropriate response from society at large. That, in turn, rests on the credibility of the science behind policy and on finding politically acceptable management solutions for the use of resources.

In this volume, we consider an emerging management tool, or approach to fisheries management, that is in contrast to long held practices where the paradigm is separation of fisheries research operations from those who fish. We are talking about a new paradigm in the case of cooperative research—partnerships for a common purpose. That purpose is sustainable fisheries. The engagement of fishers in the data-gathering process is unique, producing timely, accurate, and credible information for decision making. We have a ways to go to make this concept fully operational. However, as the discourse here makes clear, the prospects are exciting indeed.

---

[1] Current address: University of North Carolina Wilmington, Center for Marine Science, 5600 Marvin K. Moss Lane, Wilmington, North Carolina 28409, USA.

Executive Summary

Troy W. Hartley
Northeast Consortium
Department of Resource Economics and Development
University of New Hampshire
310 James Hall, Durham, New Hampshire 03824, USA

Cooperative fisheries research and management has been gaining greater attention and applications over the past 10 years. Recent national reviews and testimonies on cooperative research (NFCC 1999; Sissenwine 2001; National Research Council Committee on Cooperative Research in the National Marine Fisheries Service 2004) and cooperative management (Wilson et al. 2003; Weber and Iudicello 2004) have underscored the potential and significance of cooperative fisheries research and management in today's fisheries context. At the fourth Sea Grant-sponsored American Fisheries Society symposium, *Partnerships for a Common Purpose: Cooperative Fisheries Research and Management*, the steering committee organizers aimed to build off these reviews, to facilitate a rich discussion among fishermen, scientists, and managers that advanced our understanding about the impacts, operations, and future of cooperative fisheries research and management. The steering committee was composed of 26 representatives of the symposium sponsors that are engaged in cooperative research and/or management.

The symposium examined cooperative fisheries research and management in commercial, recreational, and subsistence fishing and fisheries management contexts, both marine and freshwater. Specifically, the symposium sought a dialog-focused meeting that

- summarized the achievements and outcomes of working cooperatively in fisheries research and management;
- cataloged the critical lessons learned from past experiences with cooperative fisheries research and management;
- identified significant challenges and future opportunities for integrating cooperative fisheries research and management; and
- advanced the fisheries research and management communities in understanding and improving cooperative research and management and considering future applications.

Panels composed of fishermen, scientists, resource managers, and project administrators presented their perspectives on discussion questions, asked questions of each other, and with the help of a facilitator, engaged audience members around the panel topics. Extensive preparatory materials, including project participant interviews, and keynote speaker and panelist white papers were compiled and used to guide discussion in the panel sessions. Keynote speakers were asked to address specific questions in their remarks and challenge the symposium participants while actively participating in the subsequent discussions. A symposium-specific poster session and reception was hosted in and around the symposium venue to promote further networking and a continuation of the discussion that was sparked within the symposium panels.

These proceedings reflect only a portion of the accomplishments at the symposium, summarizing the content of the preparatory studies, symposium discussions, and poster sessions. However, the future opportunities that emerged from the discussions, personal networking, and articulation of outstanding research questions and needs will ultimately attest to the success of this symposium. The executive summary synthesizes the impacts cooperative fisheries research and management have had for management, science, and people; the lessons learned and practitioner tips for conducting effective cooperative fisheries research and management; and finally, what future opportunities, needs, and issues exist that are central to promoting the integration of cooperative fisheries research and management. However, it is critical to define the cooperative fisheries research and management concepts, or at least make the scope and boundaries clear. What are and what are not cooperative fisheries research and cooperative management?

What Are Cooperative Fisheries Research and Cooperative Fisheries Management?

Cooperative fisheries research and cooperative fisheries management are not the same thing and have consequently often been conducted and studied as mutually exclusive phenomena. There are examples of cooperative research programs that are independent of management and cooperative management programs without research functions. While they have substantial content differences, they also share many common characteristics relating to the impacts, operations, and future potential of working cooperatively.

The symposium did not draw a sharp delineation in the definitions of cooperative research and cooperative management, but rather accepted a continuum of features that more generally bound the concepts. At some point, further subdivision and analysis may be constructive, but at this point, cooperative fisheries research and management cast broad nets and encompass many diverse examples.

Cooperative research is not a new concept in the United States and has been used widely in other countries as well. Cooperative fisheries research was commonplace prior to the expansion of the federal fisheries regulatory regime, although it has expanded greatly in the last decade (see Hartley and Robinson 2006). The practice puts scientists on board fishing vessels to gain from the knowledge of fishermen and to utilize fishing boats as research platforms, combining the scientific community's methods and credibility with the fishing community's knowledge of the marine ecosystem and marine operations. Little has been published in the peer-reviewed literature on cooperative research; however, cooperative research entails

- *industry engagement* in scientific research, although level of involvement varies;
- a variety of *scientific partners*, including government, academic, and nonprofit scientists. For their part, fisheries managers are less likely to be directly involved in the research activities, although they are increasingly serving advisory roles on research projects;
- *formal partnerships*, although the structure of partnerships can vary (e.g., number of partners, payment arrangements, level of formality in partnership agreement);
- a management or industry driver to initiate or create the *incentive* for the research;
- clear, specific *objectives* to the partnership, although scope can vary; and
- findings of *practical value* to management, industry, or other stakeholder groups.

Cooperative management, commonly referred to as *comanagement,* also has a long history both in the United States and internationally. More published empirical research on comanagement exists than on cooperative research, although there is not a single, accepted definition among comanagement scholars. Nonetheless, a few common elements include

- *formal partnership* between government and user-groups;
- *shared authority* among the partners to make resource management decisions;
- *varying scope* from a specific user- or industry-group comanagement arrangement with government, through multiple interest-based stakeholder, to broader geographic community-based stakeholder comanagement; and
- *scale* for management ranges from local to regional, although some multinational examples exist around a shared geographic area (e.g., Great Lakes).

Based upon these broad definitional spectra, the symposium explored themes and asked questions across all types of cooperative research and comanagement. The findings shed light on the similarities and differences between cooperative research and comanagement and lead to many future research questions and opportunities.

Impacts: Achievements in Management, Science, and among People

Management, science, and people experienced accomplishments, achievements, or impacts from cooperative fisheries research and management. In most cases, the symposium captured what impacts participants believed resulted from their cooperative activities, rather than what might have actually been accomplished. Table 1 summarizes perceived impacts and achievements, as described in the same words used by participants in the case studies or symposium.

Cooperative fisheries research and management can achieve a wide range of things from specific, tangible outcomes, such as answers to research questions or regulatory tools of management, to human

TABLE 1. What do cooperative fisheries research and management accomplish?

Management impacts and achievements—particularly from cooperative management

- Produces *tangible outcomes*, solving problems, expanding knowledge, and producing efficiencies, including the following:
    - Creates more expansive, holistic management view of the problem.
    - Generates broad data sets that provide ability to employ and test adaptive management strategies.
    - Builds network for ecosystem-based management.
    - Produces greater range of management options, including more shared responsibility among parties, and additional tools and practices to reduce bycatch and fish mortality.
    - Results in better understanding of fishermen perspectives and actual fishing practices.
    - Leads to productive, healthy fish stocks and protected and restored ecosystem health.
- Improves *decision-making climate* among fishermen, scientists, and managers.
    - Improves decision-making climate, less contentious dialog.
- Effective problem-solving with consensus agreements and less litigation.
    - Reduces need for enforcement, promoting greater compliance and greater support for actions.
    - Enhanced coordination and consolidation of management actions.

Scientific impacts and achievements—particularly from cooperative research

- Produces *tangible outcomes,* understanding, capacity, and stock assessments.
    - Achieves specific scientific research goals, including answers to specific basic and applied research questions and long-term data sets.
    - Enhances research capacity, financial efficiency, coordination.
    - Enhances understanding of science and fishermen's knowledge.
    - Increases interdisciplinary research and expands scope of research questions.
    - Expands knowledge base, including greater scientific knowledge and inclusion of local knowledge.
    - Increases understanding of stock structure, including additional stock assessments and bycatch and fish mortality levels.
- Improves *scientific administration and output.*
    - Enhances coordination and consolidation of research.
    - Provides greater support for scientific findings.

Human impacts and achievements

- Expands *industry involvement,* particularly in scientific research.
    - Expands industry capacity and involvement.
    - Promotes broader cooperation and understanding between industry sectors.
    - Supports alternative income sources for fishing communities.
    - Provides venue for additional stakeholder involvement (e.g., conservation organizations, recreational fishing sector).
    - Empowers fishermen.
- Builds *trust and respect* between fishermen, scientists, and managers.
    - Improves mutual understanding, mutual respect, and trust.
    - Builds common language.
    - Builds partnership capacity.
- Facilitates *broader ecosystem perspectives* and *shared responsibility* for the future.
    - Promotes a sense of stewardship among resource users.
    - Shares and integrates scientific and fishing industry knowledge.
    - Promotes new ideas, creativity.
    - Provides opportunities to bridge diverse professional cultures.
- Generates *hope, optimism, and excitement* for the future.

dimensions relating to the professional working relationships among fishermen, scientists, managers, and other partners. These achievements are not reached easily, and symposium participants chronicled many lessons learned for how to make cooperative fisheries research and management work effectively.

Lessons Learned: How-to's of Cooperative Research and Comanagement

Cooperative fisheries research and management are not entirely new ideas. Examples of community-based comanagement are more prevalent overseas than in the United States, although the existing regional fishery management councils are a comanagement system on a regional scale. Cooperation in fisheries research and management can be extremely difficult. The stakes are high, and the science and research are debated in an open public forum. Table 2 below summarize the factors that initiate and administer cooperative fisheries research or management operations, the challenges each face, and the underlying motives that keep participants engaged over the long duration.

TABLE 2. How do cooperative fisheries research and management work?

Getting cooperative fisheries research or management started

- New programmatic *opportunities* and enhanced funding for fishermen and scientist to partner.
- *Reforms* in management regime and institutions, particularly for comanagement opportunities.
- *Preplanning* in scoping meetings and research planning that identified needs and set priorities, particularly for cooperative fisheries research.

Managing operations

- *Engage partners* early and often, and if necessary, expand stakeholder group and participant engagement during implementation.
    - o Provide wide range of options for participating (i.e., multiple role and responsibilities available for prospective participants).
- Establish *clear, relevant, and agreed-upon objectives.*
- Maintain *flexible approaches* to achieve clear goals.
- Ensure adequate *time management and planning,* often requiring more time than expected.
- Ensure adequate *staff capacity* to manage people, information, and communication.
    - o Employ project coordinators, and outreach, education and data management staff.
    - o Adequate leadership.
    - o Employ sufficient coordinating mechanisms.
- Follow consistent *decision-making protocols.*
    - o Generate and apply clear operating procedures.
- Advance *open, transparent, and inclusive communication.*
    - o Employ multiple communication tools and mechanisms.
- Promote constructive professional *working relationships.*
    - o Maintain clear roles and responsibilities.
    - o Engage skilled, capable partners.
    - o Keep participants motivated and engaged.
- Be creative to leverage and obtain adequate and *consistent funding.*
- Maintain high *professional quality standards.*
    - o Monitor progress.
    - o Adhere to the latest safety, insurance, contracts, and compensation standards and practices.
    - o Conduct independent, peer review, audits, and other external review mechanisms.

Challenges during implementation

- Inconsistent funding *undermines commitment* to partnership.
- *Integrating fishermen's knowledge* with the scientific or management framework.
    - o Obtaining acceptance and buy-in from affected parties.
    - o Finding feasible solutions that meet party needs.
    - o Overcoming divergent interests.

TABLE 2. Continued.

- Reconciling *data sharing and data control* (e.g., publication opportunities), particularly in cooperative research.
- Adequate *management and coordination.*
    - o Managing the volume of information and data.
    - o Coordinating all the parties.
    - o Securing adequate time to fully participate.
    - o Securing capable partners.

Sustaining cooperation by addressing participants' interests and needs to

- leverage resources and promote *efficiencies*, including meeting funding and employment needs.
- enhance *likelihood* of behavior change and genuine impact.
- respond to new *expectations* in today's fisheries research and management.
- produce greater synergistic benefits and *multidisciplinary capacity*, including engaging industry, and integrating fishermen and scientific knowledge.
- build *mutual understanding and trust.*
- promote an individual's *alternative* viewpoint or perceived need for an integrated management strategy, including concerns with existing knowledge or management system.
- *learn* more about the resource, particularly among cooperative research participants.
- advance an individual's *sense of duty and responsibility.*

Cooperative fisheries research and management are not the same thing. One keynote speaker called research "sweetness and light," while viewing management as a "brutal process, tension and fight, grey clouds." Nor is there a single blueprint or model for each. They change as the context changes, maturing and constantly adjusting over time. With change comes conflict, and many of the lessons learned were learned the hard way. Nonetheless, symposium participants were optimistic about the future of cooperative fisheries research and management, believing that challenges could be overcome and that slow, incremental change in society and governance were moving cooperative fisheries research and management significantly ahead.

Future: Learning More About Cooperative Fisheries Research and Management

The future of cooperative fisheries research and management looks bright with expanded opportunities, additional focus on integration of the research findings into management, and concrete benefits for the professional working relationships among fishermen, scientists, and managers. While new challenges have emerged, there was a genuine optimism among symposium participants for what lies ahead. Table 3 summarizes some of the key issues faced by cooperative fisheries research and management as greater integration of cooperative research and management occurs.

There is no single strategy for integrating cooperative fisheries research and management. Integration is specific to the political, managerial, and scientific context. Trust, credibility, and previous success contribute to defining the political, management, and scientific context. The future of cooperative fisheries research and management will depend on building effective partnerships; conducting credible, relevant, and high-quality science; integrating research results into management decisions; and a real-time feedback loop to management and the research community about emerging needs and what worked and what did not. Despite the challenges, the symposium participants were enthusiastic about the future, identifying many areas where greater cooperation could prove beneficial:

Topics ripe for cooperative fisheries research

- Stock assessment, including better real-time monitoring of mixed stock complexes.
- Monitoring and observing.

TABLE 3. Promoting integration of cooperative fisheries research and management.

Expansion of stakeholder participation

- Encourage involvement of conservation organizations, recreational fishing industry, shoreside businesses, and other stakeholder groups.
- Consider who should participate and responsibilities for promoting participation.
- Develop adequate communication and coordination tools to promote broad partnerships.
- Promote broad menu of options for participation (e.g., varying degrees of responsibilities and and engagement).

Consistent funding

- Maintain the success of current cooperative fisheries research and management efforts through adequate funding.
- Demonstrate genuine long-term commitment to partnerships by ensuring adequate funding.
- Promote broad collaborative, multiparty funding commitments.

Importance of independent third parties that complement the National Marine Fisheries Service (NMFS)

- Independent cooperative research funders, data managers, peer reviewers, and project coordinators are needed to ensure credibility of research and address broader, holistic, and longer-term ecosystem questions.
- Independent programs can more easily address broader ecosystem issues and thus complement NMFS cooperative fisheries research and management, which often have more immediate, management-driven needs.
- Third party intermediaries (e.g., Sea Grant) provide important translating functions between parties, and conduct outreach and education that bring research findings back to industry participants.
- Liaisons are needed to facilitate the integration of research findings into the management process.

Improved understanding of the human dimensions

- Understand the factors that contribute to building institutional and individual trust, credibility and legitimacy in the science and management decisions, integration of scientific and fishermen's local knowledge, and transparency in the decision-makingprocess.
- Advance knowledge of strategies to adequately address power inequities, conflicts of interests, and competing interests and needs.
- Measure actual socioeconomic and ecological benefits accrued through cooperative fisheries research and management.
- Study fishing and compliance practices and the affect of regulation on fishing practices and markets.
- Consider strategies for overcoming communication and professional cultural differences among diverse partners.
- Understand the role of communication in building mutual respect and understanding and effective partnerships.
- Study whether involving partners in research empowers them and increases their involvement in management.

Integration and coordination in governance

- Promote high research standards and quality, and a rigorous, multistakeholder process of research priority setting.
- Conduct broad spectrum of research from monitoring and observing systems, through applied research, to long-term strategic research and hypothesis-driven research.
- Facilitate adequate communication among independent funders to optimize use of limited resources, while promoting innovation and meeting stakeholder needs.
- Incorporate greater outreach and education into cooperative research projects to close the loop, fully informing industry, scientists, and managers about findings.
- Advance governance mechanisms for fairly addressing power inequities, conflicts of interests, and competing interests and needs.

TABLE 3. Continued.

Ecosystem-based management
• Integrate science and societal values in decision-making procedures, including providing mechanisms to resolve disputes among parties.
• Apply tools that are appropriate for the scale and can integrate across scales for broader analysis or management.
• Build adaptive research and management protocols that supply real-time feedback on outcomes of decisions to managers and on emerging research needs to scientists.

- Gear development and habitat impact studies.
- Ecosystem analysis.
- Model development and testing for adaptive management.
- Socioeconomic impacts on fishermen, communities, and markets.
- Decision-making process, including participation in allocation of harvest.

Topics ripe for cooperative fisheries management

- Ecosystem-based management.
- Opportunities for experimental pilot projects within current U.S. governance system of laws and regulations.

Finally, as opportunities arise to conduct comanagement experiments on smaller scales in the United States (e.g., area and sector management), a commitment is needed to conduct science on the same scale and build the real-time feedback loops into the research and management systems. Consequently, some of the best case studies of integrating cooperative fisheries research and management may emerge in the design and implementation of area and sector management pilots currently underway throughout the United States. The lessons these comanagement experiments may provide could contribute significantly to facilitating the integration of cooperative fisheries research and management on broader regional scales.

Additional details on these findings can be found in the remainder of these proceedings. Part I summarizes the preparatory case study materials. Part II contains the symposium papers and discussion summaries. Part III chronicles the poster abstracts, which were accompanied by a reception during the symposium to facilitate further discussion.

Last, the steering committee organizers and sponsors of the fourth Sea Grant-sponsored American Fisheries Society symposium, *Partnerships for a Common Purpose: Cooperative Fisheries Research and Management*, were tremendously grateful for the time, commitment, and contributions of all case study and symposium participants, more than 150 in all. These proceedings and the symposium success were a testament to their genuine commitment and belief in fisheries partnerships for a common purpose.

References

Hartley, T. W., and R. A. Robertson. 2006. Emergence of multi-stakeholder driven cooperative research in the northwest Atlantic: the case of the northeast consortium. Marine Policy 30:580–592.

NFCC (National Fisheries Conservation Center). 1999. Investigating cooperative data gathering. Technical report. National Fisheries Conservation Center, Ojai, California.

National Research Council Committee on Cooperative Research in the National Marine Fisheries Service. 2004. Cooperative research in the National Marine Fisheries Service. National Academies Press, Washington, D.C.

Sissenwine, M. 2001. Fisheries cooperative research. Testimony of Michael Sissenwine, Director of Northeast Fisheries Science Center before the Committee on Resources, Subcommittee on Fisheries Conservation, Wildlife and Oceans, U.S. House of Representatives . Available: http://www.nefsc.noaa.gov/misc-docs/Sissenwine-Final-Testimony.pdf (May 2006).

Weber, M. L., and S. Iudicello. 2005. Obstacles and opportunities for community-based fisheries management in the United States. Coastal Enterprises, Inc., Wiscasset, Maine.

Wilson, D. C., J. R. Nielsen, and P. Degnbol, editor. 2003.The fisheries co-management experience: accomplishments, challenges and prospects. Kluwer Academic Publishers, London.

Symbols and Abbreviations

The following symbols and abbreviations may be found in this book without definition. Also undefined are standard mathematical and statistical symbols given in most dictionaries.

A	ampere
AC	alternating current
Bq	becquerel
C	coulomb
°C	degrees Celsius
cal	calorie
cd	candela
cm	centimeter
Co.	Company
Corp.	Corporation
cov	covariance
DC	direct current; District of Columbia
D	dextro (as a prefix)
d	day
d	dextrorotatory
df	degrees of freedom
dL	deciliter
E	east
E	expected value
e	base of natural logarithm (2.71828...)
e.g.	(exempli gratia) for example
eq	equivalent
et al.	(et alii) and others
etc.	et cetera
eV	electron volt
F	filial generation; Farad
°F	degrees Fahrenheit
fc	footcandle (0.0929 lx)
ft	foot (30.5 cm)
ft^3/s	cubic feet per second (0.0283 m^3/s)
g	gram
G	giga (10^9, as a prefix)
gal	gallon (3.79 L)
Gy	gray
h	hour
ha	hectare (2.47 acres)
hp	horsepower (746 W)
Hz	hertz
in	inch (2.54 cm)
Inc.	Incorporated
i.e.	(id est) that is
IU	international unit
J	joule
K	Kelvin (degrees above absolute zero)
k	kilo (10^3, as a prefix)
kg	kilogram
km	kilometer
l	levorotatory
L	levo (as a prefix)
L	liter (0.264 gal, 1.06 qt)
lb	pound (0.454 kg, 454g)
lm	lumen
log	logarithm
Ltd.	Limited
M	mega (10^6, as a prefix); molar (as a suffix or by itself)
m	meter (as a suffix or by itself); milli (10^{23}, as a prefix)
mi	mile (1.61 km)
min	minute
mol	mole
N	normal (for chemistry); north (for geography); newton
N	sample size
NS	not significant
n	ploidy; nanno (10^{29}, as a prefix)
o	ortho (as a chemical prefix)
oz	ounce (28.4 g)
P	probability
p	para (as a chemical prefix)
p	pico (10^{212}, as a prefix)
Pa	pascal
pH	negative log of hydrogen ion activity
ppm	parts per million
qt	quart (0.946 L)
R	multiple correlation or regression coefficient
r	simple correlation or regression coefficient
rad	radian
S	siemens (for electrical conductance); south (for geography)
SD	standard deviation
SE	standard error
s	second
T	tesla

tris	tris(hydroxymethyl)-aminomethane (a buffer)
UK	United Kingdom
U.S.	United States (adjective)
USA	United States of America (noun)
V	volt
V, Var	variance (population)
var	variance (sample)
W	watt (for power); west (for geography)
Wb	weber
yd	yard (0.914 m, 91.4 cm)
α	probability of type I error (false rejection of null hypothesis)
β	probability of type II error (false acceptance of null hypothesis)
Ω	ohm
μ	micro (10^{26}, as a prefix)
‘	minute (angular)
“	second (angular)
°	degree (temperature as a prefix, angular as a suffix)
%	per cent (per hundred)
‰	per mille (per thousand)

Part I: Setting the Stage

Editors' Note:

The preparatory materials aimed to frame the deliberation at the symposium sessions and program. Taylor Singer, Kearney, and Fetterolf first lay the groundwork for what constitutes cooperative research and comanagement. They highlight that current literature contains a definitional continuum across common factors that reflect the spectrum of both cooperative research and comanagement. While greater precision may be beneficial to future research on these phenomena, Taylor Singer's, Kearney's and Fetterolf's objectives were to frame the definitional questions and provide some thought-provoking suggestions for what predicts success in cooperative research and comanagement. Hartley and Read report on the findings from interviews and surveys with 42 practitioners. They too are seeking to frame the topics and guide deliberations rather than reach definitive empirical conclusions. Practitioners reported specific achievements from their collaborations; lessons about initiating, managing, and sustaining collaboration; and concrete tips for others undertaking cooperative research or management efforts.

In summary, this section sets the stage for the symposium speakers, panelists, and audience participants. It also provides a helpful "how-to" guide for initiating, operating, and sustaining cooperative research and comanagement initiatives. Many of the tips also form the basis for testable hypotheses in future empirical study of collaboration in research and management.

The authors of the background materials compiled a short list of references from the existing literature on cooperative research and comanagement. To date, more has been published on comanagement than cooperative research, although cooperative research is gaining increasing attention from scholars and several empirical studies are currently underway.

Berkes, F., R. Mahon, P. McConney, R. Pollnac, and R. Pomeroy. 2001. Managing small-scale fisheries: alternative directions and methods. International Development Research Centre, Ottawa.

Borrini-Feyerabend, G., M. Pimbert, M. T. Farvar, A. Kothari, and Y. Renard. 2004. Sharing power: learning-by-doing in co-management of natural resources throughout the world. International Institute for Environment and Development, London and International Union for the Conservation of Nature, Commission on Environmental, Economic and Social Policy, Tehran, Iran.

Hanna, S. 2003. The future of fisheries co-management. Pages 309–319 *in* D. C. Wilson, J. R. Nielsen, and P. Degnbol, editors. 2003.The fisheries co-management experience: accomplishments, challenges and prospects. Kluwer Academic Publishers, London.

Jentoft, S. 2003. Co-management – the way forward. Pages 1–14 *in* D. C. Wilson, J. R. Nielsen, and P. Degnbol, editors. 2003.The fisheries co-management experience: accomplishments, challenges and prospects. Kluwer Academic Publishers, London.

National Research Council, Committee on Cooperative Research in the National Marine Fisheries Service. 2004. Cooperative research in the National Marine Fisheries Service. National Academies Press, Washington, D.C.

Pinkerton, E. 1992. Translating rights into management practice: overcoming barriers to the exercise of co-management. Human Organization 51(4):330-341.

Pinkerton, E. 2003. Toward specificity in complexity: understanding co-management from a social science perspective. Pages 61–77 *in* D. C. Wilson, J. R. Nielsen, and P. Degnbol, editors. 2003.The fisheries co-management experience: accomplishments, challenges and prospects. Kluwer Academic Publishers, London.

Weber, M. L., and S. Iudicello Martley. 2004. Executive summary of obstacles and opportunities for community-based fisheries management in the United States. A report to the Ford Foundation. Coastal Enterprises, Inc., Wiscasset, Maine.

American Fisheries Society Symposium 52:3–6, 2006

Scope of Cooperative Fisheries Research in the United States

Laura Taylor Singer
Gulf of Maine Research Institute
350 Commercial Street, Portland, Maine 04101, USA

The 21st century has brought with it new and revitalized efforts to conduct fisheries research that involves scientists and fishermen as research partners. Cooperative research has become a common practice throughout the United States, and many areas of the country are looking for ways to increase the use of this approach. This practice puts scientists on board fishing vessels to gain from the knowledge of fishermen and to utilize fishing boats as research platforms. The combination of the scientific community's method and credibility with the fishing community's knowledge of the marine ecosystem and marine operations has dramatically increased the quality and quantity of information available to support marine resource management decisions.

Cooperative research is not a new concept in the United States and has been used widely in other countries as well. An early example comes from the Southeast Fisheries Science Center in Miami where, beginning in 1954, recreational and commercial fishermen were used to tag sailfish *Istiophorus platypterus*, blue marlin *Makaira nigricans*, white marlin *Tetrapturus albidus*, swordfish *Xiphias gladius*, bluefin tuna *Thunnus thynnus*, and yellowtail tuna *T. albacares*. Alaska fishing vessels have been used as partners on research projects since the 1950s and later used in surveys and gear research. In recent years, the National Marine Fisheries Service (NMFS) has expanded the number of projects where commercial and recreational vessels have been used as platforms. In New England and the West Coast, there are now formal cooperative research initiatives underway to promote the involvement of scientists and fishermen in both developing and implementing research projects. The National Research Council (NRC) completed a review of cooperative research efforts in 2004 to document the success of these projects and explore effective design and implementation strategies. The NRC noted that there are few studies of cooperative research published in the peer-reviewed literature and a precise definition does not exist. The NRC used the following working definition of "cooperative and collaborative research" (2004:8)

> The nature and level of cooperation can vary greatly among projects. At one end of the spectrum are projects with relatively low levels of cooperation, such as NMFS chartering commercial vessels for surveys (in which the primary form of cooperation is commercial crews helping in the actual daily operations of the surveys) or fishermen keeping logs of fishing activities. On the other end of the scale are cooperative research projects where fishermen and agency personnel work together in all phases of the project, including development of the research question design of the project, performance of research, analysis and interpretation of results, and communication and dissemination of study findings. These types of projects are often referred to as "collaborative research."

Given the years of experience and the more recent expansion of cooperative research programs, the question remains, what is cooperative or collaborative research and what is not.

What Is Cooperative or Collaborative Research?

Several factors appear important in characterizing cooperative/collaborative research; each will be discussed in greater detail below, followed by some short examples:

- Industry engagement in scientific research, although level of involvement varies.
- Variety of scientific partners, including government, academic, and nonprofit scientists.
- Can, but do not necessarily, directly involve fisheries managers as research partners. For their part, fisheries managers are less likely to be directly involved in the research activities, although they are increasingly serving advisory roles on research projects.

- Formal partnerships, although the structure of partnerships can vary (e.g., number of partners, payment arrangements, level of formality in partnership agreement).
- Management or industry driver to initiate or create the incentive for the research.
- Clear, specific objectives to the partnership, although scope can vary.
- Findings of practical value to management, industry, or other stakeholder groups.

The *level of involvement* of the fishing industry in cooperative research projects varies greatly. Some projects have already been designed and funded before fishing vessels are recruited to serve as research platforms. This form of cooperation was typical in the early surveys conducted by NMFS and provided a cost-effective approach to fisheries research and supplemental income for fishing vessels. At the other extreme, fishermen and scientists develop an idea from inception, often in response to a management or industry needs, and work together to secure funding and execute the project. This form of relationship is often referred to as "collaborative research," as it encompasses a more equal partnership among parties throughout all the phases of the project.

There are several examples from which to draw upon to illustrate the range of cooperative relationships that have been formed among scientific, fishing, and management communities involved in fisheries research. These *partnerships* include participants from across a spectrum of academic, government (state, federal, and tribal), and nonprofit scientists. In many cases, commercial and recreational fishermen are involved, either as individuals or through organized associations. Many of the cooperative projects involve the management community. The projects are often in response to a management need or results are anticipated to have a direct management impact. Throughout the country, federal funding has been instrumental in catalyzing several cooperative research projects and creating an administrative structure with which to support these efforts. Industry too has built organizational capacity to manage and conduct cooperative research activities in their professional associations.

The number of vessels participating in cooperative research projects varies from a single vessel to nearly 100 in the case of some tagging studies. A range of *industry compensation* methods has been utilized in projects. These included exempting volunteer vessels from trip limits while engaged in cooperative research and providing monetary compensation through daily charter rates for vessels. Other methods of compensation incorporated the sale of catch to offset the vessel charter costs or provided free observer coverage (which is usually paid for by the vessel) in lieu of direct monetary compensation. In all cases, the participants work together to collect the data. However, the analysis is most often conducted by the scientific partner with consultation with the other members of the project.

West Coast Groundfish Mesh Size Study

One of the more well-known cooperative research projects on the West Coast involved several partners as part of an advisory group to oversee the design of the experiment, to monitor the implementation phase, and to advocate for the dissemination of the findings. The advisory committee included members of industry, federal and state scientists, academics, and fisheries managers. The West Coast Groundfish Mesh Size Study was deemed successful due in large part to the guidance of the advisory committee and the *clear objectives* established early in the process to guide the experimental design. In this case, although lead by a core research team, the involvement of an advisory committee consisting of broad stakeholders and the use of industry vessels provided a collaborative partnership among all parties. The results of the study revealed that an increase in the groundfish mesh size would reduce the number of unmarketable fish discarded. In 1991, the Pacific Fisheries Management Council voted to increase the mesh size in part due to the additional data provided by this research project.

Experimental northern shrimp fishery in the Gulf of Maine

Many successful cooperative research projects involve a fishing industry association or fishermen interested in pursuing a particular *applied research question*. In those cases, they seek out the technical expertise of the scientific community and jointly raise funds to execute the project. There are several gear research projects that have been initiated through this type of relationship. For example, Associated Fisheries of Maine and the Gulf of Maine Research Institute secured funding in 2001 from the NMFS Northeast Regional Office to conduct an experimental shrimp fishery on Cashes Ledge in the Gulf of Maine. The goal was to test whether shrimp could be harvested during May and June around Cashes Ledge (a traditional fishing ground 60 mi offshore in the Gulf of Maine) using the Nordmore grate. The Nordmore grate hangs inside the net and was designed with 1-in bar spacing to allow groundfish to escape while retaining shrimp. The challenge was to catch shrimp while protecting juvenile groundfish in order to support the recovery of groundfish stocks in the Gulf of Maine.

Six Maine-based fishing vessels from Boothbay Harbor, Cundy's Harbor, Portland, and Saco were involved in the project. The vessels were able to harvest shrimp and document less than 3.5% bycatch (by weight) of groundfish. The research was used to inform the shrimp management process. Two years later, the New England Fishery Management Council reopened the Cashes Ledge area for a spring shrimp fishery as part of Amendment 13 to the Multispecies Fisheries Management Plan.

Great Lakes Fishery Commission

Expressions of concern for Great Lakes fisheries were voiced by Canada and the United States as early as 1830. But efforts to establish international commissions and/or complementary regulations and management programs failed repeatedly from 1893 to 1952. However, by 1946, the sea lamprey *Petromyzon marinus*, native to the Atlantic Ocean and a parasitic predator that feeds on the body fluids of fish, had invaded the Great Lakes and was recognized as an impending disaster for fisheries. This threat provided an added *incentive* to recast and complete earlier negotiations, and the Convention on Great Lakes Fisheries between Canada and the United States was entered into force in 1955. The convention charged the commission to control sea lamprey, to develop and coordinate fishery research and management, and to advise governments on measures to improve the fisheries.

The commission pursues much of its program through a complex committee structure involving many scientific disciplines and the agencies with mandates for fisheries and sea lamprey research and management. The Council of Lake Committees is made up of senior staff members from each agency administering a Great Lakes fishery assisted by experts and advisors from all agencies concerned. From the early 1980s, the commission has steadily increased its budget for research. Since that time, the commission has funded and received 83 fishery and 140 sea lamprey research completion reports. This research has been accomplished through contracts with the U.S. Geological Survey's Hammond Bay Biological Station and by soliciting research from several other scientists and institutions.

Further Characteristics: Factors Affecting Success

The NRC report identified the following factors as affecting the success of cooperative research projects (2004):

- Scientists and fishermen realize that each bring valuable tools and experience.
- All participants understand that they are involved in scientific research and value scientific methods.
- Research is collaborative (i.e., more than cooperative).
- Parties determine that working together is the most effective means to solve a problem.
- Involvement of project coordinators, leaders, and/or advisory groups.

At the same time, the NRC acknowledges that different approaches have been successful for different types of problems and in different institutional settings.

The NRC concluded that fishermen are *motivated* to participate by many factors, including the financial incentives, desire to investigate alternative hypotheses or views about the status of the resource, or an interest in improving the scientific knowledge underlying management. Scientists are motivated by financial incentives, opportunities to publish, chance to work on a problem with real and immediate impact, and time on the water away from the lab and office.

Regardless of the incentives, both fishermen and scientist partners must be *committed* to working together to successfully orchestrate a cooperative research project. Early and regular *communication* is most often the key component affecting success. Cooperative research takes advantage of the expertise of fishermen and scientists, often with varying viewpoints about natural resources. Both parties must be willing to take the additional time needed to *learn from each other*, be open minded, and alter preconceived ideas to generate the best product. All parties should value each other's role in the process and also acknowledge the limitations of their experience. Many projects find using a *project coordinator* valuable to organize the project participants, keep the goals focused and defined, and communicate among all parties. These individuals should understand the importance of the scientific research being conducted and its potential utility to the management and fishing communities. They should feel equally comfortable talking with the fishermen partners as well as the scientists.

In order to be successful, cooperative research should maintain the same rigorous *scientific standard* as those expected in other research projects. Scientists should be accountable for the science, and the results should be peer reviewed. Fishermen, as well, should operate their vessels in a highly *professional standard*, adhering to the latest safety standards, possessing adequate insurance, and formalizing expectations and compensation in contractual agreements.

Finally, the NRC report highlighted the need to communicate the results of cooperative research. It is essential that the results be disseminated in a timely fashion after being vetted through a peer-review process. Many cooperative research projects are initiated by a need among managers for additional information or a desire by fishermen to augment the data being used to support management decisions. It is important for the data to be analyzed and peer reviewed prior to dissemination. It is equally important that the review process be timely so the results can be made available to all the partners in the project and become a useful source of new information.

American Fisheries Society Symposium 52:7–10, 2006

Scope of Cooperative Fisheries Management Globally

John Kearney
5064 Doctor's Brook, Rural Route 3
Antigonish, Nova Scotia B2G 2L1, Canada

Carlos Fetterolf
697 Old Forge Court, Chelsea, Michigan 48118, USA

For the past few decades, fisheries management has been based on the view that fishing is highly competitive and can easily lead to overfishing and a resulting depletion of fish stocks and loss of economic rent. Thus, fisheries management has often been characterized by command and control regulations of fishing operations by government managers guided by scientific information and methods. However, the implementation of these rules has led to a complicated regulatory process that is costly and difficult to enforce.

Recent trends in the natural and social sciences and concerns among stakeholder groups raise questions about the effectiveness of this conventional form of management. First, there is an increasing recognition of the complexity of aquatic ecosystems and the need to incorporate an understanding of ecological processes into fisheries management. This will likely require more adaptive management that is capable of responding to uncertainty and change. Second, there is an appreciation of the value of incorporating other kinds of knowledge, in addition to scientific knowledge, into management systems to help deal with this increasing level of complexity. Third, there is recognition of the need to make better use of the human capital of the fishery by bringing fishers and other stakeholders into the management arena through their involvement in scientific and participatory research, the design of management plans, and monitoring and enforcement.

All of these factors have started people thinking about a kind of fisheries management that is decentralized and on a smaller scale, more focused on local habitats and fish stocks, and that spreads the responsibility and cost of management among a wider segment of the population. Thus arises the theory and practice of fisheries cooperative management, more commonly referred to as comanagement.

What is Cooperative Management or Comanagement?

There is no single definition of comanagement. Table 1 below illustrates four different definitions drawn from the comanagement literature.

Three out of the four definitions include all of the following elements in some combination: comanagement is a *partnership* between government and user-groups in which *authority is shared* in making resource management decisions.

The definitions are in less agreement about who should be part of this partnership. Most definitions include government as one of the partners. From the literature, one can identify three broad categories of comanagement based on who is included as partners with government.

- *User-based or industry-based comanagement.* An industry or user organization(s) comanages with government. Example: management of a groundfish stock over part of or its entire range by a joint industry/government board.
- *Multistakeholder comanagement.* A wide variety of organizations representing a broad array of interests are comanagers with government. Example: management of an anadromous fish stock and its habitat in a watershed.
- *Community-based comanagement.* All the stakeholders in a geographical area have an internal process for reaching management decisions and deal collectively with the government. Some defi-

TABLE 1. Definitions of comanagement from the literature.

Jentoft (2003): Comanagement is a collaborative and participatory process of regulatory decision-making between user-groups, government agencies, research institutions, and other stakeholders. Power sharing and partnership are an essential part of this definition.

Berkes et al. (2001) Fisheries comanagement can bedefined as a partnership in which government, the community of local resource users (fishers), external agents (nongovernmental organizations, academic, and research institutions), and other fisheries and coastal resource stakeholders (boat owners, fish traders, money lenders, tourism establishments, etc.) share the responsibility and authority for making decision about the management of a fishery.

Borrini-Feyerabend et al. (2004): A partnership by which two or more relevant social actors collectively negotiate, agree upon, guarantee and implement a fair share of management functions, benefits and responsibilities for a particular territory, area or set of natural resources

Pinkerton (1992): Comanagement can be generally defined as power sharing in the exercise of resource management between a government agency and a community or organization of stakeholders.

nitions of this kind of comanagement involve no partnership with government at all or devolution of authority by the government to the community. Example: First Nation or tribal management of all the fisheries in their territory.

Pinkerton (2003) describes some of the characteristics of *mature comanagement* through an analysis of one of North America's oldest comanagement arrangements, the Washington State treaty tribes comanagement arising from Judge Boldt's decision in 1974. Thus, in mature comanagement, one finds that

- government is viewed as an ally in the comanagement partnership.
- management activities go beyond sustainable harvest management to include other functions such as allocation, habitat protection, and policy making.
- power sharing is scaled up and down the levels of decision making to the level most capable of successful operation.
- comanagement stimulates broader participatory democracy in civil society.
- comanagement is based more on collective rights of a group than on individual rights.
- sustainable comanagement includes some control over the marketing of fish.
- the fishing community is able to capture the benefits of its own stewardship.

Mississippi Interstate Cooperative Resource Association (MICRA)

MICRA, the Mississippi Interstate Cooperative Resource Association, was formed in 1991 by the Mississippi River basin states in response to the rapid decline of paddlefish *Polyodon spathula* in the 1980s. This species was once abundant in the drainage and provided robust commercial and recreational opportunities. The states recognized the need to manage the paddlefish on a multistate basis. They joined forces to develop a common approach and share financial and manpower resources in the river basin that drains 41% of the continental United States.

MICRA has since evolved to address issues involving other aquatic organisms (e.g., Asian carp, aquatic nuisance species, and native mussels). MICRA now also addresses river management issues such as floodplain ecosystems, the aquatic nuisance barrier in Chicago, levees and floodplain management, navigation, power generation, and reservoir and channelization projects. MICRA's mission has become to "improve the conservation, development, management and utilization of interjurisdictional fishery resources in the Mississippi River Basin through improved coordination and communication among the responsible management entities."

To date, 28 of the basin's state fish and wildlife agencies have signed on, making it one of the country's largest resource partnering efforts. The states have been joined by several federal and tribal entities.

Scotia-Fundy Community Management Boards

The Scotia-Fundy Community Management Boards are an example of comanagement initiated and implemented at the grassroots, community level.

In 1996, Canadian fishing organizations, representing some 2,000 fixed-gear (hook-and-line and gill-net) groundfish license holders along the Atlantic Coast of Nova Scotia and New Brunswick, formed 10 community management boards. Their actions were sparked by declining groundfish stocks and cutbacks in federal government services to the fishing industry. Their goal was to manage the groundfishery to meet community economic and social objectives within the context of local ecological conditions and within the conservation framework set out by government managers. Lacking enabling legislation for such boards, the fishing organizations established their own legal framework by requiring license holders to sign a contract to abide by the board's management plan and to accept sanctions for violations of that plan. Each license holder had the choice to fish under a federal government management plan or a community management plan. Eventually, more than 98% of all active license holders chose the community plans.

The community management boards have been able to optimize the benefits to be derived from the groundfishery under conditions of resource scarcity while maintaining a broad distribution of those benefits in the fishing community. The boards have also fostered increased participation by fishing organizations in cooperative research with government and university scientists. Some boards, convinced of the benefits of community-based management, have also initiated and become involved in community economic development through the formation of marine resource centers.

Further Characteristics: Factors Affecting Success

Based on the work of Berkes et al. (2001), Jentoft (2003), and Hanna (2003), the following factors have been identified as affecting the success of comanagement:

- *Motivation.* Comanagement has a greater chance of success if it is founded on the authentic desire to effectively manage fish stocks based on notions of participation and shared stewardship.
- *Institutional integration.* Trusted institutions are established to foster group cohesion, inclusivity, transparency, leadership development, and accountability.
- *Appropriate scale.* For each fish stock, comanagers must find a scale of management that is large enough to encompass important ecosystem variables and small enough to be effective in representing fisheries participants. For some, the scale of decision making should be guided by the subsidiarity principle: management decisions should always be made at the lowest possible level. It should start as the level of the individual or firm and, as necessary, scale up to the community, regional, and national levels.
- *Knowledge base.* A challenge of comanagement is how to integrate large-scale scientific knowledge with small-scale, local knowledge. Under comanagement, this challenge is met by broadening the knowledge production process from a small group of specialists to all fisheries participants.
- *Participation and representation.* Successful comanagement is more likely produced when fishery participants and managers are empowered by improving their skills in management activities, social relations, and decision-making processes through "learning by doing," education and training, and membership in effective organizations.
- *External support.* Comanagement is facilitated by the existence of laws that enable the effective organization of fishers and other stakeholders and that provide clear lines of authority and responsibility for the comanagement process. Similarly, research institutions and a variety of

nongovernmental organizations can provide crucial independent advice to fishery managers and participants.

- *Management performance*. Within a comanagement setting, an evaluation of fisheries management must include an assessment of biological, ecological, economic, and social outcomes and the comanagement process itself. This evaluation should be guided by the establishment of indicators of success that are agreed to by all participants before implementation.

In closing, comanagement examples vary considerably across scales and in their degree of maturity. As comanagement is implemented, there appears to be an increased level of partnership with government and overall power-sharing, and expanded involvement in related issues such as allocation, habitat protection, policy making, and marketing of fish.

References

Berkes, F., R. Mahon, P. McConney, R. Pollnac, and R. Pomeroy. 2001. Managing small-scale fisheries: alternative directions and methods. International Development Research Centre, Ottawa.

Borrini-Feyerabend, G., M. Pimbert, M. T. Farvar, A. Kothari, and Y. Renard. 2004. Sharing power: learning-by-doing in co-management of natural resources throughout the world. International Institute for Environment and Development, London and International Union for the Conservation of Nature, Commission on Environmental, Economic and Social Policy, Tehran, Iran.

Hanna, S. 2003. The future of fisheries co-management. Pages 309–319 *in* D. C. Wilson, J. R. Nielsen, and P. Degnbol, editors. 2003.The fisheries co-management experience: accomplishments, challenges and prospects. Kluwer Academic Publishers, London.

Jentoft, S. 2003. Co-management – the way forward. Pages 1–14 *in* D. C. Wilson, J. R. Nielsen, and P. Degnbol, editors. 2003.The fisheries co-management experience: accomplishments, challenges and prospects. Kluwer Academic Publishers, London.

National Research Council, Committee on Cooperative Research in the National Marine Fisheries Service. 2004. Cooperative research in the National Marine Fisheries Service. National Academies Press, Washington, D.C.

Pinkerton, E. 1992. Translating rights into management practice: overcoming barriers to the exercise of co-management. Human Organization 51(4):330-341.

Pinkerton, E. 2003. Toward specificity in complexity: understanding co-management from a social science perspective. Pages 61–77 *in* D. C. Wilson, J. R. Nielsen, and P. Degnbol, editors. 2003.The fisheries co-management experience: accomplishments, challenges and prospects. Kluwer Academic Publishers, London.

American Fisheries Society Symposium 52:11–28, 2006

Testimonies from Fisheries Managers, Scientists, and Industry: Achievements, Lessons, and Advice

TROY W. HARTLEY
Northeast Consortium
Department of Resource Economics and Development
University of New Hampshire
310 James Hall, Durham, New Hampshire 03824, USA

ALESIA N. READ
American Fisheries Society
5410 Grosvenor Lane, Suite 110, Bethesda, Maryland 20814, USA

A wide variety of cooperative fisheries research and management initiatives are underway throughout the United States and around the world. What have we learned from these examples, and how can we move the science and practice of cooperative fisheries research and management forward? Testimonies were gathered from participants in successful cooperative fisheries research and management examples worldwide to prepare panelists and audience participants for the fourth Sea Grant-sponsored American Fisheries Society symposium *Partnerships for a Common Purpose: Cooperative Fisheries Research and Management.*

Methods

Thirty-seven cases and key case participants were nominated by the symposium steering committee members. Based upon their knowledge of the field, the steering committee felt these cases were innovative and demonstrated important lessons on how to do cooperative research and management well. See Appendix A for the complete list and short descriptions of the 26 cooperative research and 11 cooperative-management cases, along with a breakdown of geographic distribution and scale of cases and stakeholder groups of the study participants. While a geographically diverse sample was sought, there are more cases from the northeast (35% of sample) and Pacific Northwest/Alaska (22%) than other regions (southeast Atlantic, 11%; Great Lakes, 8%; Canadian maritime, 8%; mid-Atlantic, 5%; non-U.S./non-Canadian, 5%; Gulf of Mexico, 3%; and U.S. nationwide, 3%). This was due in part to the case participants that responded to the solicitation, the geographic composition and case suggestions of the symposium steering committee members, as well as the recent expansion of cooperative research activities in the northeast, Pacific Northwest, and Alaska. The cases varied in size, complexity, and scale, ranging from local cases and issues, through regional, state, multistate/provincial, national, to multinational examples.

While some small cooperative projects had very few partners (e.g., one scientist and one or two fishermen on a research project) and focused on a very specific, local question, other research and management cases contained hundreds of formal partners and hundreds more informal or occasional collaborators and volunteers. The breadth of partner types was considerable:

- Commercial fishing industry
- Recreational fishing sector
- Shoreside support businesses
- Local governments (e.g., planning departments, town councils)
- State government agencies/departments (e.g., marine, natural resource, fish and game)
- U.S. federal agencies (e.g., National Marine Fisheries Service science centers, U.S. Geological Survey, U.S. Fish & Wildlife Service, National Park Service, U.S. Bureau of Land Management, U.S. Army Corps of Engineers)
- Regional governance councils, commissions, interagency groups, and multinational commissions

- Canadian agencies (e.g., Department of Fisheries and Oceans)
- Funding agencies (e.g., government, foundations, nongovernmental)
- University partners (e.g., scientists, Sea Grant)
- Private sector consultants and business communities
- Environmental nongovernmental organizations
- Community nongovernmental organizations
- Citizen volunteers

Each key participant was contacted by email and/or phone, soliciting their participation in the study. Among the individuals expressing an interest and willingness to participate, phone or in-person interviews were conducted with 15 participants in these projects, and using the same set of questions (see Appendix B), 27 responded to a survey. Forty-two participants from each stakeholder group—fisheries managers (8), scientists (16), fishermen or other industry representatives (11), and other project participants, such as project coordinators (7)—were interviewed or surveyed, 31 from cooperative research cases and 11 from comanagement examples.

All interviews and surveys were completed between February and April 2005. In some cases, participants wanted to share more and supplied a wide array of case documentation (e.g., technical reports, educational materials, strategic plans, annual reports, etc.). In addition, Internet searches were conducted with available search software and Web-based case information gathered. Interviews, survey responses, and documentation were reviewed, coded for content, and analyzed for patterns among codes (e.g., frequencies of concepts mentioned, clustering of concepts mentioned by certain stakeholder groups, etc.) consistent with standard content analysis methods (Miles and Huberman 1994; Lofland and Lofland 1995).

The summary of results below is *not* an all-inclusive list of critical factors and lessons learned in making cooperative research or management work, but rather a complete list of issues considered important by the participants interviewed or surveyed. Given the small sample sizes and the limited duration of data collection on cases that often were many months or years in duration, the data need to be interpreted cautiously. This is particularly true for factors identified by respondents in either cooperative fisheries research or management, but not both. We cannot say that the factor is not at play in both, only that it was not raised by cooperative fisheries research *and* cooperative-management participants in this sample.

What Was Achieved by Working Cooperatively?

Participants were asked whether and why their projects were successful and what they found personally rewarding from participating in cooperative fisheries research or management. What they said resulted in a long list of accomplishments that included tangible outcomes and critical improvements in the relationships among stakeholders and the decision-making climate. In most cases, these measures captured what impacts participants believed resulted from their cooperative activities, rather than what might have actually been accomplished. While cooperative fisheries research and management did share a few similar types of achievements, each had unique accomplishments as well.

Participants identified a wide range of tangible research and management outcomes, as well as improvements in the professional working relationships between managers, scientists, and fishermen. While the scale of the accomplishments differed between cooperative research and management (e.g., answering a specific research question about stock structure versus generating more effective management strategies to promote healthy stocks), participants from both research and management cited the same sorts of tangible and professional working relationship accomplishments. One slight difference was the greater regularity in which cooperative fisheries research participants mentioned enhancing their understanding of their partner's knowledge (i.e., science or fishermen's knowledge) and expanding industry involvement in projects. These ideas were not raised by the 11 cooperative-management participants in this study.

Overall, participants cited improved capabilities, effectiveness, and efficiencies to solve problems or test research questions and the production of concrete agreements and valuable scientific data. They improved their professional working relationship through building trust and mutual respect, including enhanced mutual understanding of science and fishing practices, enhanced partner coordination and efficiency, and expanded industry involvement. Furthermore, participants cited an increased sense of hope, optimism, and excitement for the future, including healthier fish stocks, broader perspectives on management, sense of stewardship, and an improved decision-making climate.

Table 1 summarizes the accomplishments cited by participants, including specific participant testimonies relating to those accomplishments.

Table 1. What did working cooperatively accomplish?

Achievements:	Selected participant testimonies:
Producing *tangible outcomes*, solving problems, expanding knowledge, and producing efficiencies.	"Without the collaborative work, it is likely some of those questions may never have been answered, let alone asked."—Government research scientist, cooperative research. "Collecting information in an area where no other information is available, but which has the highest landings of the whole stock complex. Detecting spawning grounds and aggregations in new and historical locations."—Project coordinator, cooperative research. "Saving U.S. dollars with volunteer sampling and new, cost-effective instrument designs."—Government research scientist, cooperative research. "We've made progress in restoring depleted stocks."—Government manager, cooperative management. "Helping Commissioners and our staff solve problems; watching them them accomplish more than they thought they were capable of doing." — Government manager, cooperative management.
Enhancing *understanding* of science and fishermen's knowledge.	"Prior to this project, I had no faith in statistical surveying, as many commercial fishermen do. . . .After the surveys were completed I still had some doubt if this data was correct. I researched and found a 1994 state survey and the data was identical to our data. From that day on I have worked on several surveys and stand behind the data completely." —Commercial fisherman, cooperative research. "Learning more from fishermen."—Government research scientist, cooperative research. "The fishermen are very knowledgeable."—Government research scientist, cooperative research.
Expanding *industry involvement*, particularly in scientific research.	"The involvement of so many people in the fishing industry, it is amazing!" —Industry representative, cooperative research. "Everyone works well together and shares ideas and they are very flexible. This established relationship is very successful. This project has spawned many new projects."—University research scientist, cooperative research. "Lots of members of the industry were involved in collaborative research for the first time."—Nonprofit research scientist, cooperative research.
Building *trust and respect* between fishermen, scientists and managers.	"Building trust in working with the industry cooperatively."—Government research scientist, cooperative research. "[We] have built a lot of trust."—Industry representative, cooperative research.

Table 1. Continued.

Achievements:	Selected participant testimonies:
	"It is most rewarding to be truly working with professionals in a [fishing] business who are absolutely dedicated to this resource. I have nothing but admiration for the people who manage the fishery. So it is a group of people that I respect tremendously."—Project coordinator, cooperative management.
Improving *decision-making climate* among fishermen, scientists, and managers.	"The Board of fishermen ... have been very keen on not having our partnership become adversarial. They want everything that happens to prepare our responses together so that we always have a unified front even though we may have internal disagreement."—Senior government manager, cooperative research. "Very good cooperation between all parties, each group gave what they had and had a positive attitude about getting things done."—Government research scientist, cooperative research.
Facilitating *broader ecosystem perspectives* and *shared responsibility* for the future.	"Our greatest accomplishment has been to get the states to begin communicating more effectively and to view the management of interjurisdictional rivers in the Mississippi River basin as an ecosystem problem that stretches far beyond their borders, and one that we must work together on to address."—Government manager, cooperative management. "The project has been successful in holding partners accountable for how they manage the shared resource.... In a larger sense, this cooperation helps the partners to recognize the ecosystem. Through this process, the parochial boundaries are less important than natural boundaries." —Project coordinator, cooperative management.
Generating *hope, optimism,* and *excitement* for the future.	"When you see people in the industry that are making a difference, and then they talk with other people about what they are doing and they are excited about it."—Nonprofit research scientist, cooperative research. "When you go to the joint meetings you get a chance to meet and collaborate with your colleagues, you gain a lot of momentum and there is an air of hopefulness and a can-do attitude toward managinga shared resource."—Project coordinator, cooperative management.

What Lessons Were Learned About Initiating, Managing, and Sustaining Collaboration?

Participants were asked a series of questions about how their projects got started, what were their objectives, and how were the projects managed, including strategies for maintaining communication and motivation of participants. There were both similarities and differences (mostly in scale) between cooperative research and management.

Getting started

Both cooperative research and management required some initiating spark or driver to get started, although they varied greatly in what those triggers were. Cooperative research participants spoke of responding to a funding announcement and new initiatives that foster fishermen–scientist partnerships. However, prior to the development of a particular research proposal for funding, there were often extensive preproposal planning and research prioritization discussions. Getting cooperative-management initiatives off the ground required more institutional capacity and partners than cooperative research. Often institutional and management reforms were needed to permit greater integration and coordination in fisheries management. Cooperative management could not get started without these reforms, but the reforms manifested themselves differently depending in part upon the scale of the cooperative-management initiative. Table 2 summarizes the factors important to getting cooperative fisheries research or management started.

Table 2. How does cooperative research or management get started?

Conditions:	Important for Coop res	Important for Comgt	Selected participant testimonies:
Enhanced funding and fishermen and scientist to partner.	√		"A port liaison project was developed to involve individual new *opportunities* for fisheries in research with academia as well as government. As part of the port liaison project, local leaders in several West Coast communities have been selected to facilitate the matching of fishers with research....The Center, in partnership with the Pacific States Marine Fisheries Commission, also created a program to fund industry initiated research proposals."—Government manager, cooperative research. "The Northeast Consortium was established in part in response to the industry's call for more involvement in fisheries science and research. We administer two Requests for Proposals per year and operate an extensive outreach program to help match fishermen and cientists and to help fishermen develop their ideas into fundable research proposals."—University project coordinator, cooperative research.
Reforms in management regime and institutions.		√	"The Great Lakes use the Joint Strategic Plan to gain consensus on management of fisheries resources around the Great Lakes."—Project coordinator. "Commission was formed through a vompact ratified by Congress in 1942."—Manager. "Clam populations were high, but the fishery was closed. Approximately one hundred local clammers were shut out of the fishery. They began to organize themselves." —University extension staff.
Preplanning in scoping meetings and research planning.	√		"A series of local meetings were held to choose the project and develop the protocols for the work."—Commercial fisherman. "...a neutral forum for herring fishermen, processors, scientists and managers to develop research priorities for the Gulf of Maine herring fishery. This effort produced a consensus about research needs, established a foundation for collaboration among the herring industry and state and federal herring scientists, and set the stage for a variety of entrepreneurial research institutions to seek and secure funding to address herring research priorities in the Gulf of Maine."—Project coordinator, nongovernmental organization.

Goals and objectives

Having been successful in getting cooperative projects initiated, participants spoke of significant administrative, communication, and personnel management responsibilities. It starts with clear goals and objectives for both cooperative research and management. While cooperative research participants cited very specific scientific research objectives and activities associated with their experimental designs and hypotheses, almost all participants identified a second objective—to build partnership capacity for proactive engagement in solving fisheries problems. In their own words, participants said, for example,

> "[Our goal was to] see if a new kind of net design would be more selective, but the other goal was to involve the fishing industry in gathering data, which would allow them to empower the fishing industry to work collaboratively."—Nonprofit organization research scientist.

> "An important goal of this project was to find out if a community could plan, fund, and execute a research project that would collect data that agencies would then use for management decisions."—Project coordinator.

Participants in cooperative management generally identified a broader range of objectives than their cooperative research counterparts; however, they were predominantly administrative in nature. Many singled out a broad or specific dimension of ecosystem health as their primary objective (e.g.,

Tips from Experienced Practitioners: Getting Started

Practitioners underscored how important it was to get started on the right foot. Getting partnerships off the ground meant simultaneously addressing many diverse relational factors. As one senior government manager involved in cooperative research put it, "There needs to be money, initial trust, genuine honesty, and you have to identify some people who are committed to making it work because it is extremely time consuming." In fact, experienced practitioners, when asked what they would do differently, often cited the need to foster more industry and management involvement.

Engage partners early

"When you decide to do a project, you need to be involved right from the beginning… You want to have everyone involved in the planning so that a lot of stuff can be ironed out before you have already written the full proposal, and you have worked it out together. Everyone works together from the beginning."—Commercial fishermen.

"Need time with partners up front to gain clarity on all issues before we start. This works to minimize misunderstandings along the way."—Project coordinator.

Expanding involvement

"Needs to be lots of collaboration on the front end of the project to empower the fishermen and keep them informed that this research work is not against them, but working for them."—Commercial fishermen.

"Do the research on more of a local scale, have it run by local fishermen's associations and give the money to each community to do the work and then pool the research together. It could be done if you had good standardized training. There would be more ownership and it would be done faster."—University project coordinator.

"Push for earlier input from NMFS regarding data analysis needs and expectations for how this data can be used in management."—Nonprofit organization research scientist.

health of habitat, wildlife, or fish stocks). They were conducting a variety of activities toward this end (e.g., recovery/restoration, species population management, direct funding, monitor system health, rule-making, and enforcement). In addition, participants cited coordination as an objective as they sought to facilitate ample communication, less duplication of effort, greater efficiency, and coordinated implementation of projects and funding. Finally, several fishermen noted that community control of the rules and their future was an important goal for their grassroots, community-based comanagement efforts. Participants said,

> "[We] strive to improve information bases and data management, communication and coordination, and species and habitat management, while eliminating duplication of effort and achieving overall cost savings to taxpayers and user groups."—Government manager.
>
> "[We aim] to facilitate wise management decisions and sustainable management practices under community-based management."—University research scientist.
>
> "To be able to acquire more allocation and obtain more authority over decision-making… we felt that if our community was to survive then we had to be somewhat in control of certain things that would affect us."—Commercial fisherman.

Managing collaboration

The cases examined varied in size and complexity from two partners—one fisherman and one scientist—to hundreds of partners, with paid staff, multiple governmental agencies and scientific institu-

Tips from Experienced Practitioners: Goals and Objectives

Practitioners emphasized the importance of clear, relevant, and agreed-upon objectives, while exhibiting sufficient flexibility in implementation to meet the interests and needs of the partners. Clarity and flexibility were not at odds with one another, but rather contributed directly to ensuring successful outcomes in the complex and dynamic social, economic, and environmental context of fisheries research and management.

Clear, relevant, and agreed-upon objectives

"I think it is absolutely crucial that participants share a common vision for the project… People get involved in projects for all sorts of reasons, and the goal of the project can become clouded or forgotten. Clearly defining the goals of a project is vital."—State cooperative research scientist.

"Make sure what you are working on is relevant and important."—Cooperative research project coordinator.

"Probably the single largest issue is developing jointly acknowledged objectives and bringing all of the parties' objectives to the table for evaluation."—Government manager, cooperative management.

Flexible approaches to achieve clear goals

"The fishermen for this project were able to let the other partners know up front that there were areas where we needed to be flexible in the design of the work."—Commercial fishermen.

"Understanding how the fishermen log data themselves and adjusting to their practices is necessary. For example, a scientist cannot expect to get data in metric units when most fishermen deal with lorans, fathoms, and Fahrenheit."—Government research scientist.

tions, commercial and recreational fishing industry and shore-side support businesses, volunteers, and many others. Nonetheless, we found the nuts-and-bolts of managing cooperative fisheries research or cooperative-management initiatives, whether small or large scale, to be complicated and similar in nature. Participants in both cooperative research and management spoke of specific administration, communication, and motivational needs and strategies. They cited the need for flexibility in implementation, while maintaining clear, concise goals, objectives, and partner roles. Clear roles for each participant established accountability and eliminated confusion and possible miscommunication. Participants spoke of continuous engagement with their partners and the demonstration of genuine commitment on the part of all parties.

The *administrative* systems included project coordinators, financial management, and tracking progress. Several participants acknowledged the important role that a project coordinator could play since; as one project manager noted, "You live and die by the regularly scheduled meetings." A coordinator or facilitator managed the transaction costs of partners and meetings, coordinated tasks, activities and funding, and tracked partner responsibilities and expectations. Often, this was an individual or a dedicated staff member, although occasionally larger coordinating boards or committees augmented the individual staff member. Coordinators managed data and information sharing, maintained project histories, records of past decisions, and decision-making procedures and standards and performed other vital organizing tasks.

A few participants spoke about their funding strategies and sources—particularly the ideas of multiple sources, consistency in some level of funding, and Congressional and partner government's commitment. While both cooperative fisheries research and management participants identified funding as an important element to managing their projects, it was more acute for the research participants as they sought support project by project.

Participants assessed progress to determine whether a project was making a difference. In fact, assessing how you are doing was found to be helpful in keeping participants motivated and engaged; as one university project manager of a cooperative research project noted, "People get discouraged – constant communication about any little success really helped." Several specific activities were mentioned by both cooperative fisheries research and management participants:

- Work plans with specific milestones, benchmarks, performance metrics
- Pre- and post-meetings on project goals
- Self-evaluation
- Peer review
- Regular meetings and progress reports
- Audits

As one project coordinator concisely summarized, they use "written documents, post-meeting evaluation surveys, review comments on reports, then implement plans" in order to assess progress. A cooperative research scientist said, "Internal scientific audits were conducted, lots of participation from the industry, and fishermen who participated saw results from their work. The project was successful because the scientific goal was realized."

Almost universally, *communication* was declared the key ingredient in making cooperative research or management function effectively. The use of many communication mechanisms was important to ensure that the diversity of parties could be effectively reached. As one cooperative research project coordinator said, "Because all parties [need to] meet on an equal footing to discuss issues of common concern, multiple venues [are best] because many different kinds of audiences can be reached in this way, including academic, practitioner and bureaucratic." A fisherman added, "Very few [fishermen] have computers and some only want to deal with the issue at the table at that time." Overall, several distinct communication mechanisms were identified by the participants, but participants never used only one tool in their programs:

- Annual meetings, state-by-state meetings, and regional business meetings
- E-mail list serves

Tips from Experienced Practitioners: Managing Collaboration

The administrative activities underlying cooperative fisheries research and management can be significant. As one large-scale project coordinator summarized, "We developed a detailed schedule for plan development and adoption, and then used modeling tools to assist with the aquatic and terrestrial assessments. We had to then develop the infrastructure to implement [the plan] in the region, within each state, and within the sub-basins. We also had to develop a budget and a process for contracting, [building] a schedule and milestones, and means for coordinating this endeavor in the basin. It was a mammoth effort and took many people to make it happen."

Practitioners supplied advice on timing, staff, information and data management, decision-making procedures, and partnerships. Valuable lessons – some learned the hard way. Expect tasks and activities done collaboratively to take significantly more time than anticipated. Thus, consider hiring project coordinators, cultivating champions, building information and data management capacity, and addressing partner needs.

Time management and planning

"You need to pre-plan a lot; you need to work out all of the plans; you will need to build in time to make changes."—Government scientist.

"Don't be optimistic in your work plan. Multiply the time and expense of your first estimate by two."—Nonprofit organization research scientist.

"You need to have a good plan. You need to know what you are up against and have planned for it, before [data] starts coming in. Be patient and be realistic with the promises that you make. Set the expectations in a realistic place and people generally understand."—Industry representative.

Staff capacity

"A good coordinator or manager, who can put complex issues and problems into an understandable format "—Government manager.

"Someone has to be responsible for organizing and managing; otherwise you will not have complete agendas and productive meetings."—Project coordinator.

"Need man-power to analyze the data and get it done quickly. Working on the data is a huge effort."—Government scientist.

"Publish your data or at least make it available to everyone. Posting it on the web perhaps."—Project coordinator.

Decision-making protocols

"Need a written decision-making process and you need to be formal about it. Need to respect the formal decision-making process and be more thorough, lay out business rules. We use consensus."—Government scientist.

"Group needs a clear-cut organizational chart and constitution with by-laws by which the decision-making process is laid out and clear to all involved."—Government manager.

"Choose a decision-making process and stick with it."—Industry representative.

Promoting constructive working relationships

"Stay committed; the handshake at the very beginning is extremely important and begin there when people call with issues. Be available and accessible."—Government manager.

"If compromise can't be reached, provide for a minority position in the group position, and take the issue to each member organization for a position. Be as flexible as possible, but don't let unnecessary diversity of ideas or resources get in the way of efficiency."—Government manager.

"Be aware of groups that are sensitive, that have not had good experiences in the past."—Government scientist.

"Human beings have excellent radars for determining, especially in new situations, if their ideas and opinions are respected. I don't think respect can be effectively faked. Mutual respect creates the ability to share concerns and information, knowing it will be received. Mutual respect also reduces anxiety that the project will head in directions the participants are uncomfortable with." – State research scientist.

Funding

"Leverage support with other organizations. We tried to quantify in-kind contributions. Keeping track of in-kind participation has been helpful to get more money from other organizations."—Project coordinator.

- E-mail messages
- Phone conversations, conference calls
- Professional presentations at public meetings, council/commission meetings, fishermen advisory/steering committees
- Scoping meetings in ports, stakeholder forums
- Informational workshops, roundtable meetings
- Peer and partner reviews
- Web site
- Posters at the dock
- Word-of-mouth
- Articles in industry trade publications (e.g., *Commercial Fisheries News, National Fishermen*, association/cooperative newsletters)
- Television commercials, feature on television programs
- Final reports (widely distributed)
- Advisory committees
- Newsletters, brochures, glossy executive summary publications

In spite of acknowledged time constraints, face-to-face communication was considered the most effective means of maintaining and fostering open lines of communication. Those communication mechanisms that promoted face-to-face discussions, particularly with as many stakeholders present as possible, were favored, although the Web, e-mail, and phone were noted as more efficient. As one cooperative research coordinator noted, "Keeping lines of communication open – relying on face-to-face interactions. Relations always deteriorate when email or telephone begin to substitute for personal interaction."

Table 3 summarizes the challenges cited by participants in cooperative fisheries research and management.

Sustaining commitment to collaboration

Ensuring successful collaboration also entailed *keeping participants committed and engaged* throughout the long duration and hard work necessary to work collaboratively. This was often accomplished by demonstrating value, information sharing, quick turn around of findings, and celebrating successes.

Table 3. What challenges exist to cooperative fisheries research and management?

Conditions:	Selected participant testimonies:
Undermining commitment to partnership with inconsistent funding,	"[We] need a more stable funding source. Programs [that are] started and stopped could be damaging to relationships with the stakeholders."—Government scientist, cooperative research.
	"I have always called it the uncooperative research program with the variability in money, late arrival of money there was no way we could establish a cooperation with anybody because we never knew how much money we were going to get and if we were going to get it."—Government manager, cooperative research.
	"Better funding support would have made a huge difference in keeping the interest up among participants."—Government manager, cooperative management.
	"Travel budgets are huge obstacles and a headache. The organizations say that they are 100% behind cooperative research and management, but the travel budgets do not allow for the people to get together."—Government manager, cooperative management.
Integrating fishermen's knowledge with the scientific or management framework.	"A real crossroad involved arriving at a mutual understanding between social scientists and fishermen that was based on full agreement on the objectives, methodology and means of dissemination of the results."—Project coordinator, cooperative research.
	"The industry and the scientists will have to work very hard to reach consensus and establish similar research goals because they do not speak speak the same language."—Senior government manager, cooperative research.
	"We do have to deal with three different cultures [fishermen, scientists, managers]."—Commercial fisherman, cooperative management.
Reconciling *data sharing and control*, particularly in cooperative research.	"The government was not helpful in sharing their data. They were sharing everything that was published, but not everything is published."—University project coordinator, cooperative research.
	"It is very hard to get managers and NMFS to use the data and consider it; then once they accept it, they want you to hand over the raw data. But, since this data is owned, how do you keep some control as a researcher? Data sharing is frustrating. Need to deal with data sharing better, because it needs to get into the management decisions more quickly."—University research scientist, cooperative research.
	"Issues need to be resolved with long-range storage of the data. The university tends to be proprietary of the information until credit is given and publications are in...that hinders other people from using it and prevents it from being transferred to a centralized area... If the government is paying for this research, then the information should be made available to all the citizens."—Commercial fisherman, cooperative research.

For example, participants noted,

> "People get discouraged – constant communication about any little success really helped."—University cooperative-management project coordinator.

> "Emphasizing the small victories encourages continued participation and enthusiasm, and without those last two qualities, projects can get very difficult. Also, acknowledging progress helps those who

have invested their time and emotions know that their efforts are recognized and appreciated, and recognition and appreciation are as vital to participants as water and sunshine are to tomatoes."—State cooperative research scientist.

Participants also raised the importance of regular progress reports, scheduled update meetings, and other structural features as contributing to sustaining collaboration. Monitoring progress may have a role in sustaining collaboration through maintaining momentum and keeping participants engaged and motivated—the feedback achieves more than providing a pragmatic assessment of whether specific activities are being completed or not.

Furthermore, sustaining participation and commitment is easier when collaboration meets the *interests and needs* of the participating organizations and the individuals. When asked why they collaborate, participants in both cooperative fisheries research and management cases identified similar reasons, including greater resource efficiency and program effectiveness, enhanced synergistic benefits, and im-

Tips from Experienced Practitioners: Communication

Effective communication was at the root of many initiation, administrative, coordination, and partnership issues in cooperative fisheries research and management. Practitioners recommended openness and transparency in communication, including the use of professional outreach and communication assistance when necessary. While these experienced practitioners are involved in successful cases, even they wished they had done more to advance communication in their projects.

Open, transparent, and inclusive

"Significant time focused on communicating where you are coming from needs to be set aside for… to overcome any preconceived notions that you may hold about the other parties and that they may hold about you."—Project coordinator.

"Making all as transparent as possible."—Commercial fishermen.

"Open communication, you have to be open to what people are looking for, all stakeholders."—University research scientist.

"We should have a centralized communications committee… External communication needs to be expanded."—Project coordinator.

"Further, more frequent and more inclusive communication."—State research scientist.

Overcoming Challenges to Collaboration

These administrative and communication management activities contributed substantially to overcoming challenges to collaboration. In general, both cooperative research and management initiatives face funding constraints and information management and administrative challenges. Further, inconsistency in funding hindered the promotion of professional working relationships because it made it difficult to demonstrate genuine commitment. Participants cited challenges integrating fishermen's knowledge into either the scientific framework (i.e., hypothesis, scientific methods, and protocols) or fisheries management actions for cooperative research or management, respectively. For cooperative fisheries research in particular, information and data management challenges extended to the conflicting interests they encountered between data sharing needs and maintaining control over data use.

proved trust and understanding in the professional relationships. Collaboration met program needs by pooling resources, expertise, and funding and avoiding fragmentation and duplication of effort. Collaboration met interests in ensuring genuine impacts by increasing the likelihood that change would occur since research and management outcomes are more viable, accepted, and implementable. It met interests in expanding multidisciplinary perspectives and promoting greater respect, trust, and understanding.

In addition to these programmatic reasons, individuals cited other reasons for their personal involvement in cooperative fisheries research or management, including interests in exploring issues and alternative strategies, and a sense of duty and responsibility. Many individuals had genuine interests in learning more about the resource and managing it responsibly, including a belief that the current state of knowledge or management regime was inadequate. Table 4 below summarizes the reasons cited by participants for why they and their organizations participated in cooperative fisheries research or management.

Table 4. Why do participants collaborate on fisheries research or management?

Programmatic reasons:	Important for Coop Res	Important for Comgt	Selected participant testimonies:
Leveraging resources and promoting *efficiencies.*	√	√	"There is far more work to do than any one of us or any one of our states or agencies could ever hope to accomplish alone."—Government manager, coooperative management. "We have to collaborate with the industry because we need to bring their resources to the table; we do not get sufficient funding to do an adequate job of conducting research."—Government research scientist, cooperative research. "You get a better product. You get more done with less money."—Commercial fisherman, cooperative research. "In the course of the discussing how to address the particular challenge, a herring vessel captain suggested the logical way to solve the problem was to involve the vessels. In his words, 'We're out there every night, we live and die by our ability to find and catch catch herring, we're the first to see them spawn. We're out there We're out there anyway, why not use us?"—Nonprofit organization, research scientist.
Enhancing *likelihood* of change and genuine impact.	√	√	"You can have the best idea/process in the world, but if the affected people aren't involved at the onset and throughout the process, it won't be successful or likely implementable."—Project coordinator, cooperative management. "Fishermen really like the idea of being involved, that something they are doing matters and they see a lot of value added. So it makes them more willing to accept the decisions that may not be in, what they would consider, their best interests."—Commercial fisherman, cooperative research. "The most effective way to get support from the industry is for them to see that the research works and the data are good. If people have been working

Table 4. Continued.

Programmatic reasons:	Important for Coop Res	Important for Comgt	Selected participant testimonies:
			on the project, then changes are much more likely to happen."—Research scientist, cooperative research.
Responding to new *expectations* in today's fisheries research and management.	√	√	"Collaboration is not, in my view, a choice. It is a necessity if you are dealing with overlapping jurisdictional and jurisdictional and science mandates."—Government manager, cooperative management. "Collaborating is the right thing to do. I cannot imagine aying why people should not collaborate. We have limited resources and time to accomplish what we want to. It also helps with accountability."— Government manager, cooperative research.
Producing greater synergistic benefits and *multidisciplinary capacity.*	√	√	"We each have certain talents and knowledge to offer and by collaborating we can merge those into a much more powerful solution than if each of us tried to go it alone."—Government manager, cooperative management. "The nature of science today has changed. We are no longer individual investigators competing for limited funds for specific studies. We are asked to contribute to bigger, wide-ranging, multi-disciplinary programs, with free exchange of data."—Government research scientist, cooperative research.
Building *mutual understanding and trust.*	√	√	"[We collaborate] because it fosters good working relationships and builds trust between fishers and government agencies."—Government research scientist, scientist, cooperative research. "Collaboration produces stronger relationships and stronger communities."—University project coordinator, cooperative management. "We can each learn from each other and both groups can dispel myths about the other just by spending time together."—Commercial fisherman, cooperative research.

Individual reasons:	Important for Coop Res	Important for Comgt	Selected participant testimonies:
Promoting an *alternative* viewpoint or alternative integrated management strategy.	√	√	"I wanted to show that these fish [cod] were not evenly distributed [as currently assumed] and these fish interact with another management area."—Commercial fishermen, cooperative research. "I was trying to show other fishermen that scientists and fishermen could work together, this was a new idea at the time."—Commercial fisherman, cooperative research. "Because of the need to better manage inter-jurisdictional fishery resources – i.e., those fishery resources

Table 4. Continued.

Programmatic reasons:	Important for Coop Res	Important for Comgt	Selected participant testimonies:
			that cross or are common to two or more state boundaries and come under the shared jurisdiction of two or more governmental entities"—Government science manager, cooperative management. "There is a fear as to what parochialism on the Great Lakes would do to management that drives these partners to continue to work together."—Government project coordinator, cooperative management.
Learning more about the resource, particularly among cooperative research participants.	√		"Concern by fishermen and scientists that conventional models used for stock assessments had substantial uncertainties."—Government research scientist. "We need better data to manage our local fisheries. We want to continue to fish here for generations and that requires good science for sustainable management." —Project coordinator. "Fishermen wanted to know how temperature affects the resource."—Industry representative.
Advancing a *sense of duty and responsibility*.	√	√	"It was my duty to do so. As a community-minded person it then turned to be my responsibility for our future generation of fishermen."—Commercial fisherman, cooperative research "Simply a personal desire to preserve and protect our rivers for my own interest and that of future generations. There have been no financial incentives for me."—Government manager, cooperative management.

Conclusion

Cooperative approaches involving scientists, managers, industry, and other resource users are relatively new in natural resource research and management, although cooperative strategies have been applied to varying degrees in fisheries and other natural resource sectors for years. Examples of cooperative fisheries research sit on a continuum from cooperative to collaborative, varying by the extent of integration between partners. Some cooperative research projects charter fishing vessels for research activities. Others are more collaborative with fishermen serving equal roles to varying degrees throughout the scientific process from generation of the research questions through study design, data collection and analysis, to interpretation and communication of research findings. Cooperative fisheries management also varies considerably across scale (community-based local examples of co-management, through watershed, state, regional, national, to multi-national and multi-jurisdictional cases). In addition, as cooperative fisheries management is implemented, there may be an increased level of partnership with government and overall power-sharing, and expanded involvement in related issues such as allocation, habitat protection, policy making, and marketing of fish.

Both cooperative fisheries research and management face broad policy issues and challenges that will shape their futures. Cooperative research practitioners and scholars must examine and understand the significance of the spectrum between cooperative to collaborative partnerships, the importance of integrating results into the management process, and the conditions that promote a high scientific standard with trained data gatherers, quality control, and peer-reviewed results. Cooperative-management practitioners and scholars need to study and understand the significance of the full complexity of stock interactions and aquatic ecosystems, reconciling scientific knowledge with the local knowledge of stakeholders, inclusiveness of participation among stakeholders, and the balance of individual and community use rights to the resource.

These questions reflect important hypotheses for future research and fodder for discussion at the *Partnerships for a Common Purpose: Cooperative Fisheries Research and Management* symposium. Further topics worthy of discussion and research emerged from the study of practitioners. Below is a summary of those findings.

What did cooperative fisheries research and management accomplish?

- Produced *tangible outcomes*, solving problems, expanding knowledge, and producing efficiencies.
- Enhanced *understanding* of science and fishermen's knowledge.
- Expanded *industry involvement*, particularly in scientific research.
- Built *trust and respect* between fishermen, scientists, and managers.
- Improved *decision-making climate* among fishermen, scientists, and managers.
- Facilitated *broader ecosystem perspectives* and *shared responsibility* for the future.
- Generated *hope, optimism,* and *excitement* for the future.

How do cooperative fisheries research or management get started?

- Enhanced funding and new *opportunities* for fishermen and scientist to partner.
- *Reforms* in management regime and institutions.
- *Preplanning* in scoping meetings and research planning.

How are cooperative fisheries research and management implemented?

- *Engage partners* early and often, and if necessary, expand stakeholder group and participant engagement during implementation.
- Establish *clear, relevant, and agreed-upon objectives.*
- Maintain *flexible approaches* to achieve clear goals.
- Ensure adequate *time management and planning*, often requiring more time than expected.
- Ensure adequate *staff capacity* to manage people, information, and communication, including project coordinators, leadership, and/or project champions.
- Follow consistent *decision-making protocols.*
- Promote constructive professional *working relationships.*
- Be creative to leverage and obtain *consistent funding.*
- Advance *open, transparent, and inclusive communication.*
- Maintain high *professional quality standards.*

What challenges exist to cooperative fisheries research and management?

- *Undermining commitment* to partnership with inconsistent funding.
- *Integrating fishermen's knowledge* with the scientific or management framework.
- Reconciling *data sharing and control*, particularly in cooperative research.

Why do participants collaborate in fisheries research or management?

- Leveraging resources and promote *efficiencies.*
- Enhancing *likelihood* of change and genuine impact.
- Responding to new *expectations* in today's fisheries research and management.
- Producing greater synergistic benefits and *multidisciplinary capacity.*
- Building *mutual understanding and trust.*
- Promoting an individual's *alternative* viewpoint or perceived need for an integrated management strategy.

Tips from Experienced Practitioners: Implementation Headaches

Practitioners cautioned others to be aware of and plan for likely implementation challenges. The list will be long and the challenges will be case specific. Here are just a few experienced by these practitioners:

Cooperative research

"The range of this project is large. It spans a large geographic area, so it is tough to logistically communicate with everyone."—Industry representative.

"It is difficult to give fishermen direction, when they make up the Board that I work for."—Project coordinator.

"Requiring insurance will severely limit the number of eligible boats for this work."—Project coordinator.

"Some fishermen are better than others at catching fish. Some have better equipment and gear type licenses. So to come out with a fair balance is not always so easy."—Commercial fishermen.

"There is a lot of conflict with rules and regulations getting in the way of scientist's research."—Government research scientist.

"People get discouraged... Burn-out is a real issue and we don't have a solution for that. All of our partners find the problem of broad grassroots participation in their organizations is an intractable one."—University project coordinator.

Cooperative management

"Distance! We are geographically scattered over two-thirds of the state."—Industry representative.

"Maintaining neutrality."—University project coordinator.

"We also have a lot [of fishermen] that try to sneak in [to the area] by buying a license from another community and trying to activate it with us."—Commercial fishermen.

"Different cultures and gear types plus different fishing seasons have to be accommodated."—Commercial fishermen.

"Dealing with public officials and organizational representatives who continue to focus on the short-term at the expense of the long-term recovery of stocks."—Government manager.

- *Learning* more about the resource, particularly among cooperative research participants.
- Advancing an individual's *sense of duty and responsibility*.

While these similarities between cooperative fisheries research and management reflect important considerations for research and further discussion, the differences between the two also are noteworthy. Cooperative research was of a shorter duration, with specific scientific knowledge goals, while cooperative management had a much longer time horizon and included far more administrative coordination goals and objectives, given its role in stock health and resource allocation. They were similar in that they both shared partnership building objectives. Are the goals of research (i.e., seeking knowledge and greater understanding of the resource) more easily shared as a common interest among partners than the divisive allocation and fishing behavior change objectives of management? Are cooperative research partnerships easier than cooperative-management partnerships?

In closing, cooperative fisheries research and management in the United States and Canada remain relatively new. Further examination of the impacts, lessons learned, and future directions and opportunities are critical. The discussions at the *Partnerships for a Common Purpose: Cooperative Fisheries Research and Management* symposium, which are summarized in the next section, contribute to broadening our understanding and advancing cooperative fisheries research and management.

References

Lofland, J., and L. H. Lofland. 1995. Analyzing social settings: a guide to qualitative observations and analysis. 3rd edition. Wadsworth, Belmont, California.

Miles, M. B., and A. M. Huberman. 1994. Qualitative data analysis. 2nd edition. Sage, Thousand Oaks, California.

Part II: Symposium Deliberations

Editors' Note:

The symposium was divided into three stand-alone sessions that built upon the background materials from Part I and each other. Part II contains the summary of the symposium presentations and facilitated discussion. It starts with an exploration of where cooperative fisheries research and comanagement have been and what they have achieved, followed by a discussion of what we have learned about making each function effectively and culminating in a look to the future. Keynote speakers were invited to address overarching questions and to be provocative, challenging the symposium panelists and audience to examine critical issues and move our collective understanding of cooperative research and comanagement forward.

Mary Pete from the University of Alaska Fairbanks, and formerly the director of the Division of Subsistence with the Alaska Department of Fish and Game, welcomed the participants by underscoring the significance that fisheries resources have for communities, states, and the economy. She challenged the participants to consider ways to develop, enhance, and improve partnerships and suggested that some stakeholders may have less "stake" while others have more and thus should carry varying degrees of influence in the decision-making process.

Making a difference by working cooperatively

James W. Balsiger, the acting deputy assistant administrator for regulatory programs, provided National Marine Fisheries Service (NMFS) Director William T. Hogarth's keynote address. Dr. Balsiger's remarks and Dr. Hogarth's paper frame the NMFS perspectives and interests for cooperative fisheries research, including that NMFS "values cooperative research," and considers "cooperative research an integral part of the business of collecting data to support stock assessment and management activities." Hogarth and Balsiger acknowledged the challenges that practitioners face in integrating diverse perspectives through valid scientific inquiry and called for building capacity in all research partners to fill existing data gaps. Dr. Hogarth's paper challenged symposium participants to examine the positive and negative (e.g., mortality) impacts that cooperative research might have on the fish stocks and on fishing communities. They note the persistant tensions that arise between the need for timely contributions to management issues, on one hand, and the need to follow rigorous scientific procedures that demand time-consuming peer reviews, leveraging multiple sources of funds, and competion for limited available dollars.

Pat White, president of the Gulf of Maine Lobster Foundation, commissioner on the Atlantic States Marine Fisheries Commission, and a former commissioner of the Pew Oceans Commission, cited social, cultural, and economic impacts of working cooperatively in his keynote remarks. White discussed the professional growth that both fishermen and scientists achieve by working together, including the development of the next generation of fishermen. He challenged participants to include an important missing stakeholder group, conservation organizations who have participated little in cooperative fisheries research to date. White catalogs a broad list of social, cultural, and economic impacts of working cooperatively, including greater research and management program efficiencies, creation of a credible integrated science that improves understanding and appreciation of each other's knowledge and abilities through building a common language, enhanced climate and venue for professional growth, engagement of additional stakeholders, and promotion of visionary, creative thinking.

Ralph Rayburn, associate director and extension program leader for Texas Sea Grant, moderated and summarized the first of two panel discussions in this section that focused on the management and scientific impacts of working cooperatively. Individual panelists raised several substantive impacts in their opening remarks, including enhanced scientific understanding of ecosystems and bycatch issues with more interdisciplinary research and new, innovative research questions; greater efficiencies in governance, research coordination, and fund leveraging; and more adaptive management. Panelists noted that working cooperatively has lead to greater fishermen involvement in management, increased acceptance of implementable management actions, and an improved opportunity for integration of diverse organizational cultures (i.e., scientific, management, and industry). The subsequent facilitated discussion among panelists and audience members centered on specific management benefits (e.g., turtle excluder devises that are integrated into regulations and have improved species abundance) and their spill-over impacts (e.g., industry financial and in-kind support for protecting Kemp's ridley turtle nesting areas). A greater range of management alternatives have emerged from the new types of questions being asked that would otherwise not have been thought of. The broader data sets being collected and expanded have guided the initial discussion surrounding ecosystem-based management.

Troy Hartley, associate director of the Northeast Consortium and research assistant professor of natural resource and environmental policy at the University of New Hampshire, moderated and summarized the second panel that examined the social, cultural, and economic impacts of working cooperatively. Individual panelists acknowledged a wide range of human impacts (e.g., supplies financial benefits, stimulates innovative ways of thinking, generates trust among partners and greater regulatory compliance, and presents an opportunity to integrate fisheries-based and research-based knowledge). Panelists noted a unique and emergent characteristic of the cooperative process that is shaped by the existing network of relationships among people and raised several important factors to study about the human dimensions of working cooperatively (e.g., role of commitment, incentives, nature of participatory opportunity, and communication). The subsequent discussion converged on a few critical ideas. There was considerable discussion and no agreement about participation in cooperative research and comanagement. Specifically, who participates, are some more entitled than others, and whose responsibility is it to ensure that all stakeholders are aware that decisions are occurring? The discussion continued with an emphasis on information and data use, trust, incentives, knowledge integration, and communication.

Lessons learned: making cooperative research and management work

Ray Hilborn from the School of Aquatic and Fishery Science at the University of Washington chaired the U.S. National Research Council (NRC) panel on NMFS cooperative research. His keynote address focused on the stakeholders' and institutions' incentives for working (or not working) cooperatively. Hilborn suggests that in fisheries with long-term dedicated access, commercial fishing groups have strong financial incentives to ensure effective management and thus often fund research. He also drew attention to the organizational factors that inhibit collaboration (e.g., permitting, insurance, funding) and cited the NRC panel conclusions that scientific institutions contain inadequate incentives and often have strong disincentives for cooperative research.

Rollie Barnaby, a fisheries extension educator with New Hampshire Sea Grant and outreach coordinator for the Northeast Consortium, moderated and summarized the first of two sessions on lessons learned. In particularly, Barnaby's session examined factors that make cooperative research function. Panelists raised several factors in their opening remarks, including competitive proposal processes with common evaluation criteria; independent analysis and review of methods, results, and findings; and a clear decision-making framework to create structure for resolving differences and ensuring accountability. Panelists noted the importance of leadership that can deal with change as well as issue complexity, strong partnerships with realistic expectations, and diverse types of partnering opportunities to promote the widest possible participation. The subsequent discussion among panelists and the audience considered characteristics of successful projects, leadership, challenges, knowledge integration, program monitoring, and the future of cooperative fisheries research. Obtaining support, commitment, and willingness to work together from all parties is important, and gaining support from the fishing industry requires a solid consultation process with clearly defined goals, means to ensure that personal agendas are not obstructions to progress, and a commitment that the data collected will be shared. Industry is better prepared to partner when they have the organizational capacity and infrastructure to effectively manage and coordinate many individual fisherman.

Paula Cullenberg, program leader and coastal community development specialist with Alaska Sea Grant, moderated and summarized the second session in the lessons learned section. In their opening remarks, panelists identified several key factors that contribute to effective comanagement: sense of stewardship among partners; set of mutually agreed upon principles and goals; capacity and capacity-building to enable local involvement and provide the skills needed to participate; and the administrative capacity, including facilitation, communication, funding, and flexibility. In the subsequent discussion, symposium participants examined the general characteristics of successful comanagement, including consensus, accountability, communication and cooperation. The need for patience was stressed, as building professional relationships and cooperation take hard work and commitment. Listening was considered critical to true integration of local knowledge, as it demonstrates respect, generates questions, and promotes learning. Additional challenges include time, work load, funding, maintaining motivation, adequate communication, and ensuring engagement. Finally, the panel concluded that greater devolution of authority and broader community management may likely develop in coming years and it will continue to build local capacity to manage fisheries.

Opportunities: Future Needs, Interests, and Integration of Cooperative Research and Management

Clarence Pautzke, executive director of the North Pacific Research Board, supplied the keynote address to the final, culminating session on the future opportunities for cooperative research and management. Focusing on his work to develop a science plan for the North Pacific Research Board, he articulated the clear benefits of bringing the knowledge and skills of scientists and fishermen together (i.e., improved quality, quantity, and relevance of fisheries research). Pautzke sees cooperative research growing as funders actively seek opportunities to support cooperative research, view cooperative research as a cost-effective way to do ecosystem research and monitoring, and promote the training of the fishing industry in science and the collection of scientific data. He recommends the involvement of industry and scientists in identifying research needs, developing hypotheses, and establishing new ways to partner, particularly in the critical area of conservation engineering gear projects, such as research to reduce bycatch.

John Schwartz, extension program leader with Michigan Sea Grant, moderated the first session on linking cooperative research and management, with a focus on future partners' needs and interests. Alesia Read with the American Fisheries Society drafted the session summary. Panelists opening remarks included an acknowledgement that needs and interests are diverse and expanding; they have changed over time and will continue to do so. Thus, a single model for cooperation will be ineffective. Any perception that one partners' needs are being met at the expense of another partner will weaken the program and potentially render it ineffective. Managers are characterized by a need for jurisdictional respect, a recognition that they are to serve the public trust, including stakeholders who have no political voice or outlet, such as children and future generations. The subsequent discussion narrowed onto a few key points. Since each cooperative program takes place within its own unique regional context, a one-size-fits-all procedure or model will not work. Attention is needed to matching the appropriate partners who can work effectively together. Furthermore, all science, whether it is done cooperatively or not, must be held to the same rigorous, peer-reviewed standards in order to inform management. There is a need to close the information loop for both the expansion of knowledge and the dissemination of information. Finally, the panel concluded by identifying topics ripe for future cooperative research.

Jim Murray, director of outreach with the National Sea Grant Office, moderated and summarized the final session of the symposium on integrating cooperative research and management. Panelists' opening statements raised several key factors that need attention to enhance the integration of cooperative research into management (e.g., developing practical guidelines from research findings that link to specific management tools and options, liaisons or champions to advance research findings in the management decision-making arena, and a feedback loop from the management to the research community to guide research activities and priorities). Future partnerships will need more structure and capacity than most have today in order to supply these new feedback, monitoring, liaison, and communication functions. Further, panelists identified challenges to integration that include how to best promote a shift to more power sharing and empowerment of communities, while justly dealing with conflicts and constructively addressing scientific uncertainty. Nonetheless, the optimism of the panel and the entire symposium came through in this session. For example, a NMFS official suggested that in the future, the word "cooperative" will be dropped from cooperative research and comanagement because it will become implicit and fully integrated.

In summary, the keynote speakers, panelists' papers and presentations, and the session deliberations demonstrated that different stakeholder groups (e.g., government, commercial industry, scientists, indigenous communities, fishing communities, conservation organizations, recreational fishing interests, and others) have both shared and varied views about what cooperative fisheries research and comanagement are, how they should be conducted, and what they can and should achieve. All participants shared optimism for the future and thought the prospects of continued cooperative fisheries research and comanagement look good. As one participant said, "we cannot go back," and cooperative research and comanagement are established components of today's fisheries management regime. Cooperation may not yet be the norm of fisheries science and management, but cooperative research and comanagement are expanding and diversifying, while engendering greater influence and significance in today's fisheries. More and continued dialog about the cooperative research and comanagement phenomena, as was the goal of this symposium, is needed to make these scientific and management strategies effective and efficient in their future

American Fisheries Society Symposium 52:33–37, 2006

Welcome Address to the American Fisheries Society Symposium on Partnerships for a Common Purpose: Cooperative Fisheries Research and Management

Mary C. Pete
Kuskokwim Campus, University of Alaska Fairbanks
Post Office Box 368, Bethel, AK 99559

I am pleased to welcome you to Alaska and to this timely session. The American Fisheries Society's conference is one of the biggest professional gatherings that Alaska hosts; this is appropriate since fish and fishing is big in Alaska. I want to thank the Sea Grant program for convening this symposium on partnerships in research and management.

A few words about my background, which forms my perspective in this address, are in order. I am a cultural anthropologist with more than 20 years of applied social science research experience. Most of that experience is in documenting subsistence hunting and fishing patterns and harvest levels in order to help the Alaska Department of Fish and Game (ADF&G) and the state and federal management boards implement the state and federal subsistence priority laws. For the most recent 10 years until this past June, I directed the Division of Subsistence of the Alaska Department of Fish and Game for the state. The division is the social science research arm of the ADF&G—the only such agency in the nation. It employs mostly anthropologists, such as me, who work closely with management divisions, whose personnel are mostly biologists. As division director, I also chaired the U.S. delegation in the U.S./Canada Yukon River salmon treaty process, and once a long-term agreement was forged, I chaired the U.S. section of the bilateral panel charged with implementing the treaty. Treaty negotiations over Yukon River salmon had protracted over 16 years and were finalized in 2001. All of this experience required varying levels and types of cooperation and collaboration with constituents.

> “Partnerships in research and management are becoming common practice in certain parts of Alaska. However, cooperation has not necessarily become routine because individual players change as managers and researchers change jobs or retire and communities elect new representatives.”

Alaska Fish-Related Facts

The singular importance of fish in this state is manifested in many ways. This is a state of superlatives—biggest in land mass with the longest coastline. Statewide, subsistence, commercial, and sport fisheries effectively require important industries and cooperative research and management efforts. Economically, the state fish tax is of vital importance to communities that rely on it for government support. I want to focus in on a few Alaska fish facts to underscore the significance of fish.

Alaska is vast in land and water with many fisheries and related industries:

- There were more than 5 billion commercial fish and shellfish landed worth more than $1 billion in 2003 and generating more than $50 million in fees and taxes;
- Sport fisheries generate more than $500 million annually;

- Alaska hosts the largest single-species commercial fishery in the world (pollock *Pollachius virens*) more than 1.5 million tons in 2004;
- There are many fish-related jobs (more than 21,000 commercial permits, more than 18,000 crew member licenses, nearly 800 processors and buyers, and an average of 10,000 fish processing jobs per month);
- Alaska has the nation's largest fishing port in volume and second largest in value (Unalaska).
- Approximately 1 million salmon are harvested annually in subsistence fisheries and numerous other fish species used for food.
- The Yukon and Kuskokwim rivers each support approximately 1,500 subsistence salmon fishing families and about 800 commercial salmon permits.
- Subsistence production by families is the largest "employer" in rural Alaska.
- More than 10,000 subsistence halibut certificates issued each year to eligible individuals.

This next list of bullets focuses specifically on subsistence uses of fish and game in Alaska:

- Based on representative regional and comprehensive community harvest assessments since the early 1980s, it was determined that the annual subsistence harvest of fish and game in rural communities totaled more than 44 million pounds in the late 1990s.
- This translates to about 375 lbs per person per year, or about a pound per person per day of wild fish and game, primarily in the form of protein and fat.
- The annual replacement value of subsistence production in pounds of meat, fish and poultry is valued at approximately $220 million in 1994 terms, priced at $3 to $5 per pound. The replacement cost would obviously be more in today's economy.
- The statewide harvest composition is depicted in Figure 1 below. Fish is the largest portion of the annual composition of wild foods harvested by Alaskans—more than 60% is composed of fish and shellfish.
- The composition varies by region, with some communities relying on fish and shellfish for nearly 80% of their diet.
- The most dependent on fish are those communities along the coast and rivers of Alaska. In fact,

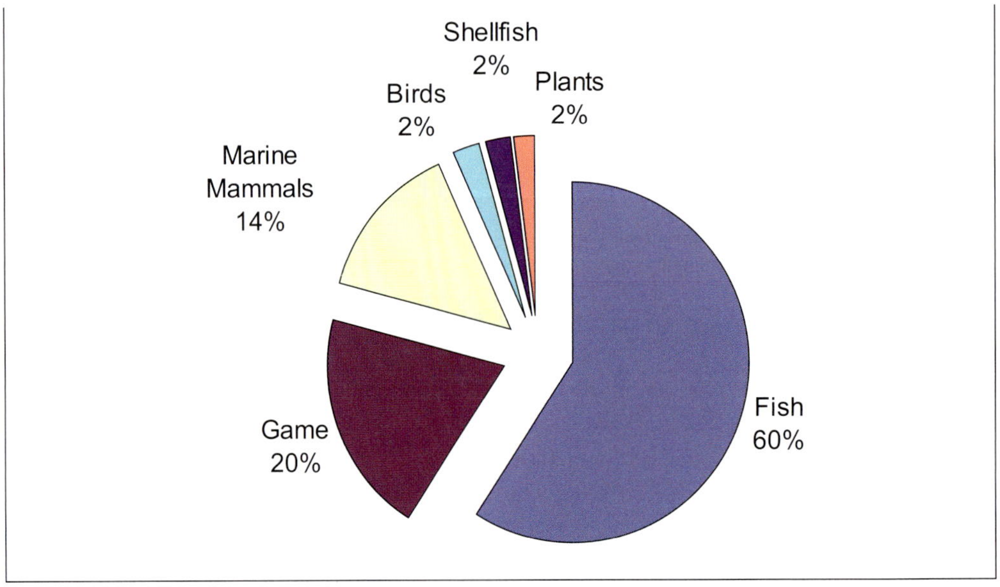

* Source: Division of Subsistence, ADF&G, Subsistence Update, 2000.

FIGURE 1. Composition of Wild Food Harvests in Rural Alaska, 1990s*

in the Yup'ik area—the Yukon-Kuskokwim delta region and the central Bering Sea coast—the word for food is also the word for fish. When you ask someone if they have eaten, you will in effect be asking if they have eaten fish.

- It is estimated that the Bering Sea is directly or indirectly the source of approximately 25 million pounds of subsistence foods annually for more than 55,000 Alaska residents, primarily Alaska natives in small, rural communities.

Cooperative Research and Management in Alaska

In my professional experience as an applied social scientist in state government, cooperation in research with participants and respondents goes without saying. In order to conduct interviews with local experts or administer harvest surveys, we needed permission from community leaders, families, and respondents. There are cases where well-planned research projects have stalled or been canceled because community leadership either denied permission or pulled support of a project. In some cases, that denial of cooperation and support is reconsidered and reversed with change in local leadership or new understanding of proposed development or regulatory impacts. Since social science focuses on human subjects, this negotiation regarding access to knowledge and action comes with the territory. One of the worst situations is to publish research findings and receive total community denouncement of the results due to lack of written prior permission and community knowledge of the research project.

There are a few basic reasons why cooperation between resource managers and researchers and people dependent on those resources in this state should and has occurred. First, the state is much too large for government to do it all with ever limited and decreasing resources. Second, increasingly, the public expects more than informed consent regarding research and management projects as interest groups realize the implications of research results and management decisions on their ability to access resources and maintain livelihoods. Last, more and more funding sources for research projects require cooperation and demonstrated efforts at capacity building. Two examples of the funding initiatives that require effective partnerships are the federal Office of Subsistence Management and the U.S./Canada Yukon River panel's Restoration and Enhancement Fund program.

> "...once agencies and communities learn to work effectively together, they tend to explore more ways to partner and build on their experience."

Partnerships in research and management are becoming common practice in certain parts of Alaska. However, cooperation has not necessarily become routine because individual players change as managers and researchers change jobs or retire and communities elect new representatives. Education and information to maintain cooperation is on-going. Further, all Alaskan partnerships that I am aware of acknowledge that the state or federal government has ultimate management authority; this is not abrogated in any cooperative agreement.

To illustrate some of the partnerships in Alaska, I will to focus on one region. In the Kuskokwim fisheries management area during 2004, there were 25 different salmon projects involving between two to six partners, with some individual partners representing more than 40 federally recognized tribes. The positive situation in the Kuskokwim area probably would not have been possible without years of formalized cooperation between ADF&G and different salmon user groups through the Kuskokwim River Salmon Management Working Group. This group was charged by the Alaska Board of Fisheries in the mid-1980s to hash out stark and seemingly irreconcilable differences between ADF&G and fishermen regarding the proposed commercial closures in the face of a limited, but fully utilized subsistence Chinook salmon *Oncorhynchus tshawytscha* resource and increasing commercial interest in that resource. This clearly shows that once agencies and communities learn to work effectively together, they tend to explore more ways to partner and build on their experience. In the Kuskokwim area, as funding opportunities developed, it became easy to imagine and realize cooperative projects such as inseason harvest monitoring (as opposed to just postseason harvest surveys), age-sex-length sampling of commercial and subsistence salmon, escapement enumeration projects such as weirs, and mark–recapture efforts. Not every project goes as well as planned, but I know that Kuskokwim area staff cannot imagine operating any other way anymore, and neither can the public they serve.

Other rural regions of the state, particularly the Yukon and Norton and Kotzebue Sound areas, have similar partnership practices and patterns. Many of these partnership opportunities grew out of funding to address salmon disasters in the late 1990s and 2000. In the case of the Yukon area, the U.S./Canada Yukon River salmon negotiations and treaty implementation require partnerships with local communities, groups, and interests.

Other than the above-mentioned reasons for cooperation, "why cooperate" is still a good question to ponder and no doubt will be well addressed in this forum. The field of resource management and research generally comes with many inherent uncertainties (e.g., the size and timing of salmon returns) that it should behoove all of us to get our working relationships in order, particularly with our constituents. Even mere informed consent generally produces some local acknowledgement of research projects and management actions—an improvement over indifference or hostility. Ideally, among many other things, effective cooperation means agreeing on a common purpose (e.g., good collection data and shared management expectations) so that bombshells are minimized.

> "All partnerships require time and effort to maintain an effective and mutually beneficial working relationship."

There are benefits that I presume everyone who has ever participated in cooperative efforts can acknowledge. Results often are more than the sum of the parts of a research or management initiative (e.g., research samples are better, dissemination of information is more effective, and young people may develop interest in resource management field). Long-term research and management partnerships in renewable resources can allow cumulative, additive work rather than reinventing the wheel each season. There has to be mutual expectation that each partner is capable of bringing their expertise or some benefit to the partnership. I think the key is to be realistic about outcomes.

Any partnerships require time and effort to maintain an effective and mutually beneficial working relationship. We need to all recognize that resource management agencies are really managing people. Based on online review of presentation titles and electronic discussions, this session assumes positive value in pursuing and maintaining partnerships. I think panel discussions need to explore and share ways to develop, enhance, and improve partnerships.

I want to pose some questions that I hope will be addressed in this symposium. What is different about management or research with partners? What makes partnerships work? Do all research and management partnerships work from the perspective of all parties? Are some research and management efforts or projects not conducive to partnering? If there are some stakeholders with less "stake," should their contribution to the partnership be less valued and therefore not as sought? Does land ownership or proximity to the resource give certain parties more leverage?

What is the future of cooperation in research and management of fisheries? The fact that there is this forum today and considerable participation in it means that there is no turning back. I think all parties contribute something to partnerships, even if it is an object lesson on how not to do things. From the Yup'ik perspective, partnerships and cooperation are essential to continued health of resources. Yup'iks and other Alaska native groups believe that if groups merely fight over a particular resource, that resource will make itself scarce and inaccessible. So, human cooperation is viewed as integral to good management and research. Cooperative efforts that do not devolve into unnecessary acrimony are viewed as worthy because they avoid offending fish and game resources, thus maintaining access and opportunities to enjoy these resources.

Again, thanks for having me present this welcome address.

Making a Difference by Working Cooperatively

American Fisheries Society Symposium 52:41–50, 2006

Principles of Cooperative Research

WILLIAM T. HOGARTH
NOAA Fisheries Service
1315 East West Highway, SSMC3
Silver Spring, Maryland 20910, USA

Cooperative research programs and projects have existed in the NOAA Fisheries Service for most of the agency's history and are considered a usual and normal approach for conducting fisheries research. This agency values cooperative research in supplementing its existing mandated and core research programs, and we have made cooperative research an integral part of the business of collecting data to support stock assessment and management activities.

In cooperative research, fishery scientists partner with commercial and recreational fishermen, the fishing industry, nongovernmental organizations, state fisheries management agencies, and universities. Partners in the regional cooperative research programs vary depending on regional research priorities and opportunities, but all programs have some level of involvement of fishermen or the fishing industry or other partnership groups in conducting regional research. It is through these partnerships that the agency gains access to knowledge, tools, techniques, skills, and experiences that are otherwise unavailable to National Oceanic and Atmospheric Administration (NOAA) scientists. This access improves the quality and completeness of our science recommendations, fosters a better understanding of NOAA science and its process, and creates more support for buy-in of subsequent management policies. Moreover, by working together, synergy is obtained not only in joint problem solving, but also in the identification of what scientific questions to undertake, their priorities, and what approaches to use through a cooperative research program.

"It is through these partnerships that the agency gains access to knowledge, tools, techniques, skills, and experiences that are otherwise unavailable to National Oceanic and Atmospheric Administration (NOAA) scientists."

The remainder of this paper identifies a set of guiding principles relevant to the future of cooperative research activities in the United States. It is my hope that these principles will be given due consideration over the course of the symposium. Additionally, this paper provides a brief background, highlights current cooperative research programs in each of the six NOAA Fisheries Service regions, discusses examples of the impact of cooperative research in each of those regions, and last, identifies future needs for those regional programs.

Principles of Cooperative Research

- Cooperative research projects must be collectively planned and executed so there is buy-in and increased understanding on all sides. By working together in partnership, scientists, stakeholders, and fishery managers can learn from each other and build mutual trust and respect. This principle supports the National Research Council's (NRC) 2004 report's recommendation on cooperative research that the NOAA Fisheries Service should have "expectations, requirements, and procedures, including the development of agreements carefully detailing the responsibilities of all participants, clarified at the beginning of every project."
- Cooperative research projects have to fit into an overall program and not be an "alternate" science with which to confront more traditional programs. Cooperative research must be part of an integrated, not separated, approach to providing information for stock assessment and fisheries management.

- Cooperative research projects must undergo the same level of peer review as traditional efforts so that there is only "one science." Information developed in cooperative research programs must not stand apart from that done by agency scientists. There are not multiple competing sciences, but a united science accounts for all data. Cooperative research results must be fully integrated into the established assessment and peer review processes. Since conditions are constantly changing in the marine environment, the period between the collection of information and management action taken should be as short as possible. This collective desire to get information into the management process quickly must not come at the cost of bypassing normal procedures for quality control/quality assurance and peer review—to do so would potentially jeopardize the utility of the information, undermine its credibility in the management process, and run the risk of the data being success fully challenged under the Information Quality Act and other statutes.
- Cooperative research projects must document more than just the final results, but also the way they are designed, developed, and implemented. Moreover, demonstrating the impacts of the final study results on the science and management processes must be captured. This reporting will serve as a major tool in providing feedback for overall program improvement.
- Cooperative research projects must be conservation neutral (e.g., no added fishing effort over and above conservation limits). Cooperative research projects vary in their size and scope and therefore impact fisheries differently. In all cooperative research programs, there is a need to accurately measure absolute fishing mortality due to research activities. By documenting the mortality of target and nontarget species, particularly stocks with low abundance, in conducting cooperative research, a full accounting of removals can be assessed.
- Cooperative research projects should be conducted in a timely manner. One facet affecting the timeliness of projects is the issuance of permits. Cooperative research participants must obtain appropriate research/fishing permit (scientific research permit, exempted fishing permit, or letter of acknowledgement) and receive approval by the appropriate NOAA Fisheries Service authority prior to initiation of the project. The need and type of research/fishing permit required to conduct cooperative research varies with the nature and scope of the cooperative research project. Moreover, additional permits may be required to comply with the Marine Mammal Protection Act (MMPA) and the Endangered Species Act (ESA). The NRC's 2004 report on cooperative research recommended that the NOAA Fisheries Service should "streamline and standardize all permitting procedures for conducting cooperative research projects so as to ensure uniform treatment and rapid processing of all applications in all regions." In response to this recommendation, the agency has been working to improve the issuance of permits, as well as the collection and handling of recordkeeping and reporting requirements for permits. Additionally, the agency is seeking mechanisms to improve the public's understanding of requirements of the National Environmental Protection Act, ESA, and the MMPA when submitting cooperative research proposals. While we have not yet streamlined all permit granting processes, significant progress has been made, and we will seek further authorities to improve these processes even more.

"Cooperative research projects must document more than just the final results, but also the way they are designed, developed, and implemented."

- Cooperative research projects need to leverage resources as much as possible. Cooperative at-sea research is an intensive use of equipment and human capital, especially when the goal for timely data delivery involves numerous platforms and groups with different capabilities. Productive partnerships and innovative approaches to offset the cost associated with research must be sought.

Background

Beginning in 1999, Congress earmarked appropriated funds for the design and implementation of cooperative research. Currently, Congressional funding supports a number of cooperative research programs, including the Northeast Consortium, the Northeast Cooperative Research Partners Initiative (CRPP), the Southeast Cooperative Research Program (CRP), the West Coast Groundfish Cooperative Research Program, and the National Cooperative Research Program (NCRP). Total Congressional funding since 1999 for national and regional cooperative research activities has approached \$85 million. In recent years, annual appropriations have totaled more than \$16 million. With this dedicated funding, NOAA Fisheries Service engages in cooperative research activities in all six of its regions with national coordination provided through the NCRP in the Office of Science and Technology.

Regional Summaries of Cooperative Research

A broad range of cooperative projects conducted throughout the United States has improved the science and management of our living marine resources. However, demands for high-quality science and more focused management continue to escalate. As future needs are identified, it is clear that the NOAA Fisheries Service must increase its efforts to leverage cooperative partnerships to maximize agency investments in science and management. Below, I summarize some of the exciting efforts that have led to improved science and management in our various regions.

Northeast region—current program

Both the Northeast Regional Office (NERO) and the Northeast Fisheries Science Center (NEFSC) are involved in extensive cooperative research activities throughout the mid-Atlantic, southern New England, Georges Bank, and the Gulf of Maine. These programs are funded by specific appropriations: NCRP, bycatch reduction funds, and special grant programs. The largest program, CRPP, is administered by the NERO and the NEFSC. The CRPP supports three long-term monitoring surveys (inshore Maine–New Hampshire, Gulf of Maine cod, and southern New England yellowtail flounder *Limanda ferruginea*), an electronic logbook study fleet (currently 30 vessels), and an extensive cod tagging initiative (>100,000 cod tagged by five cooperating organizations). Additionally, annual solicitations, request for proposals or RFPs, support shorter-term projects, including conservation engineering research to reduce incidental bycatch, evaluation of habitat and ecosystem impacts of commercial fishing, and projects to address other priority management concerns. This year's priorities include angler practices to reduce release mortality in the groundfish recreational fisheries, ecosystem-based management, and research to develop and implement strategies for enhancing safety at sea. The NOAA Fisheries Service also collaborates with the Northeast Consortium, which issues RFPs for cooperative research on a broad range of topics, including gear selectivity, fish habitat, stock assessments, and social sciences.

In the New England and mid-Atlantic regions, the New England Fishery Management Council and the Mid-Atlantic Fishery Management Council support cooperative research activities through a quota and days-at-sea set-aside program. In particular, two cooperative surveys are conducted, as are a number of bycatch reduction/conservation engineering studies. While most of the work focuses on groundfish stocks and trawl fisheries, significant work is underway with dredge, trap-pot, and hook-and-line fisheries. Commercial fishermen are most active in industry-based surveys (fishery-independent), study fleet (fishery-dependent), and conservation engineering (bycatch reduction)

"A broad range of cooperative projects conducted throughout the United States has improved the science and management of our living marine resources."

research and monitoring projects. Both commercial and recreational fishermen are very active in a number of tagging studies.

These activities are intended to not only meet agency data needs, but also to strengthen and enhance NOAA Fisheries Service's relationships with the northeast commercial and recreational fishing sectors.

Impacts of cooperative research on science and management in the northeast region

In the northeast, data from industry-based surveys will be included and add to information available for stock assessments and the evaluation of management options for a number of stocks, including yellowtail flounder and Atlantic cod *Gadus morhua.* In the case of yellowtail flounder, cooperative tagging and gear studies have helped to implement a special access program while improving the industry's understanding of agency science. With improvements in electronic logbooks and integration of vessel monitoring systems through study fleets, the level of temporal and spatial detail associated with catch observations will increase and allow finer scale and more focused management.

Future needs for cooperative research in the northeast region

Overall, there is a need for program expansion that will broaden and diversify community participation and include a greater diversity of fishery–gear combinations, vessel sizes, and sport fishermen. The longer-term goal is to assimilate and archive a rapidly growing research database and to ensure, over time, an effective means to shepherd this information into the assessment and peer-review process for eventual use in management. A second goal is to gain public and industry support to help further extend cooperative research in the mid-Atlantic, based upon the success and experiences primarily in New England.

Areas of emphasis for ongoing and future cooperative research activities in this region include

- industry-based surveys, collecting time series data on specific stocks and evaluating time area closures. There is a need for an expansion in the range of habitats being addressed through these surveys.
- study fleets and electronic reporting also hold promise for specific fleets and fisheries where the greater level of temporal and spatial detail and more timely data availability can improve precision in stock assessments.
- tagging programs conducted on a 2–3-year frequency could increase the data available for a number of stock assessments produced and provide an opportunity to engage a greater diversity of fishermen (commercial and recreational) in cooperative research and community outreach and education.
- integrated mark–recapture data management system and tag program clearing house, which will allow for long-term support for mark–recapture approaches designed to improve the accuracy and precision of analytical stock assessments.
- alternate fishing gears will complement trawl-based surveys and expand geographical coverage into habitat areas that trawls can not sample.
- bycatch reduction and conservation engineering, so that in combination with overall improvements in the temporal and spatial precision in fisheries dependent data, a greater range of management options can be analyzed.

Southeast region—current program

Cooperative research in the Southeast also has a long-standing presence and has been conducted through a host of mechanisms, including state–federal cooperative statistical programs, cooperative institutes, joint resource surveys, cooperative game fish tagging studies, and a joint industry-agency observer program. Starting in 2001, the scope of cooperative research has been broadened with the addition of dedicated funding for cooperative research projects to assist in improving confidence that both commercial and recreational fishermen have in the data and analyses performed in support of fisheries management. The enhanced program seeks to increase and improve the working relationship among researchers from the Southeast Fisheries Science Center, state fishery agencies, universities, and fishermen. As such, the principle goal of this program is to provide a means of involving commercial and recreational fishermen in the collection of fundamental fisheries information to support the development and evaluation of management and regulatory options. Projects are selected in an annual competition giving greatest weight to those projects judged to have the highest likelihood of collecting data that aid in recovering, maintaining, or improving the status of stocks; improving the understanding of factors affecting recruitment success and long-term sustainability of fisheries; and generating increased commercial value and opportunities for fishermen. Program priorities have been developed in consultation among the various regional fishery management councils, with U.S. Atlantic highly migratory species management and advisory bodies, and with regional fishing industry, scientific, and environmental constituents through a constituency workshop on southeast cooperative research.

> "[In the northeast region] Overall, there is a need for program expansion that will broaden and diversify community participation and include a greater diversity of fishery–gear combinations, vessel sizes, and sport fishermen."

The CRP serves a wide range of interests and supports fisheries management decisions of the South Atlantic, Gulf of Mexico, and Caribbean Fishery Management councils, as well as decisions regarding the U.S. Atlantic highly migratory species fisheries. In this regard, the CRP has perhaps the broadest range of interactions of any of the regional programs, particularly considering the geographical scope of the program. This broad scope is also reflected in the range of research conducted under the CRP, which has included

- development of gear modifications to reduce protected species interactions in the Atlantic pelagic longline fleet;
- evaluations of the effectiveness of bycatch reduction devices in shrimp trawls;
- shark biology and life history research;
- life history and distribution and movement studies of priority reef fish, snapper-grouper, bill fish, tunas, and other species in the region;
- on-board scientific observations of various fishing activities;
- feasibility studies for use of electronic logbooks in several fleets;
- evaluation of the efficacy of closed areas for improving stock status of fishery resources;
- cooperative fisheries statistics data collection mechanisms; and
- collection of socioeconomic information to evaluate the impacts of various regional fishery management options.

Impacts of cooperative research on science and management in the southeast region

Under the enhanced CRP, most high profile research has been on mitigating sea turtle bycatch in the U.S. Atlantic pelagic longline fisheries. Results if technologies developed under this program showed significant decreases in the bycatch of loggerhead and leatherback turtles. This encouraging performance of mitigation measures tested may be a viable solution to meet reduced bycatch requirements for longline operations not only in the Southeast, but in other regions and oceans as well. Similar cooperative research investigations have been implemented in the Pacific longline fishery off Hawaii. No less significant, however, has been enhanced life history investigations for several reef fish species (e.g., red snapper *Lutjanus campechanus*, amberjacks *Seriola* spp., red grouper *Epinephelus morio*), with results now being incorporated into stock evaluation analyses in the Gulf of Mexico and the

South Atlantic. These studies have, and will continue to provide, a more precise basis for evaluating stock productivity and potentials than previously available. As the CRP is relatively young, relatively few of the initial projects have run to completion.

Future needs for cooperative research in the southeast region

Current demand for CRP resources exceeds available funding. For this reason, refinement of research priorities will remain an important component of the overall program. While program evaluation is an ongoing process, the next major reexamination of research priorities will take place in 2006. The future direction of CRP activities in the South Atlantic and Gulf of Mexico will likely continue to revolve around research priorities identified by the councils and industry, with emphasis directed towards the collection of relative abundance patterns, catch, effort, size frequency, bycatch, and socioeconomic information in recreational and commercial fisheries. In the Caribbean, priority will likely be given to enhancing fisheries data collection mechanisms, to assessing optimal designs for marine protected areas (MPAs), to investigating the effects on recruitment related to MPAs, to addressing protected resources issues, and to habitat research, including the impacts on coral reefs from fishing activity.

> "[In the southeast region] Current demand for CRP resources exceeds available funding. For this reason, refinement of research priorities will remain an important component of the overall program."

Southwest region—current program

Cooperative research efforts in the Southwest involve commercial and recreational partners and address high priority issues for species managed by the Pacific Fisheries Management Council (PFMC). Current research focuses on groundfish, coastal pelagic species, and highly migratory species. The PFMC faces several cases of stocks falling in the "data-poor" category. Monitoring the status recovery of these stocks is often problematic because conventional sampling methods (trawling, hook and line surveys, etc.) are either not effective on hard bottom or further deplete overfished stocks. In cooperation with the Commercial Passenger Fishing Vessel fleet, new technology using acoustics and a remotely operated vehicle (ROV) is being developed to survey demersal fishes associated with hard bottom and to establish baseline biomass data, which will allow the NOAA Fisheries Service and PFMC to monitor the status and, in some cases, the recovery of overexploited stocks.

In 2001, a project was initiated with the American Fishermen's Research Foundation (AFRF) to determine movement patterns and life history strategies of North Pacific albacore *Thunnus alalunga*. Albacore are targeted by various fisheries of the North Pacific Ocean in any given year, including fleets from the United States, Japan, Taiwan, and Canada. This project is supported in part through funding provided by the AFRF, which serves as a research-based nonprofit organization spearheaded by the troll fishery and canning industries of the United States.

Impacts of cooperative research on science and management in the southwest region

Archival tagging of North Pacific albacore has provided information on migratory behavior and distribution. This long-term study involved the deployment of 500 archival tags in albacore over the period 2001–2005. To date, 277 archival tags have been deployed and 16 have been recovered. These returns represent the only recoveries of archival tags for this species in the eastern Pacific (Japanese scientists recently recaptured a fish in the western Pacific that was at liberty for nearly a year.). Archival tag data are integral to providing information on migratory behavior and distribution and will be critical to developing stock assessments. The project is now tagging at higher rates in order to derive sufficient recoveries to characterize the migration rate of the population as a whole.

Future needs for cooperative research in the southwest region

Of the 80 species managed under the groundfish fishery management plan (FMP), fewer than 20 have sufficient data to allow determination of the stock status relative to stock size and harvest rate. Addressing this assessment gap and increasing our knowledge of habitat use by groundfish populations are central goals of the region's stock assessment program. Cooperative research is an integral component of this effort. Other priorities include the establishment of a nearshore species survey, research on coastal pelagic species (e.g., market squid), and projects addressing gear interaction with protected species. Further development of acoustic/ROV survey methodology is another key focus because of the increasing use of MPAs as a management tool and because some habitats cannot be sampled well with traditional methods. Locally, this research will enable monitoring of the recovery of fish stocks in the Cowcod Conservation Area in Southern California. Nationally, the use of acoustic remote-sensing tools in combination with ROVs can overcome problems faced by traditional survey methods over hard bottom.

“[In the southwest region] Of the 80 species managed under the groundfish fishery management plan (FMP), fewer than 20 have sufficient data to allow determination of the stock status relative to stock size and harvest rate.”

Northwest region—current program

The Northwest Fisheries Science Center has conducted multiple surveys using chartered vessels to monitor groundfish populations. With the establishment of the West Coast Groundfish Cooperative Research Program in 2001, industry participation has been greatly expanded. The program's primary partners are Oregon Sea Grant and the Pacific States Marine Fisheries Commission. This multifaceted program is designed to provide opportunities through annual solicitations (RFPs) to the fishing community to work with scientists in all facets of research. Innovative approaches to cooperative research in the northwest include the recently initiated Port Liaison program, where industry leaders take part in matching individual fishermen with researchers (academic, state, or federal) to advise and assist in groundfish research along the West Coast. This assistance can range from review of proposals to the collection of samples.

Impacts of cooperative research on science and management in the northwest region

Cooperative groundfish bottom trawl surveys on the continental slope have provided invaluable fishery-independent information for several West Coast species, including Dover sole *Microstomus pacificus*, thornyheads *Sebastolobus* spp., and sablefish *Anoplopoma fimbria*. As the primary source of data for stock assessment models, this information has been critical in developing scientific advice for setting catch limits for these species. Additionally, industry is now partnering with federal scientists to develop new surveys for overfished widow rockfish *Sebastes entomelas* and canary rockfish *S. pinniger*. The industry is involved in all phases of these new surveys from planning, through design and execution.

Future needs for cooperative research in the northwest region

For the northwest, future cooperative research needs will continue to focus on the expansion of groundfish surveys both in number and geographic scope. Cooperative research efforts aimed at developing new technologies to assess life histories for overfished species, improving bycatch reduction methods, assessing fishing impacts on habitat, and generating more robust recreational data collection methods are other major needs.

Alaska region—current program

The Alaska Fisheries Science Center (AFSC) has a long history of cooperation with various groups in the Alaska fishing industry to carry out a broad range of studies. These studies have focused on improving quantitative assessments of crab and groundfish stocks, reducing bycatch in trawl fisheries, reducing the incidental take of marine birds, improving the assessment data base for sablefish and rockfish, and evaluating the effects of mobile fishing gear on bottom habitat. Industry involvement has been at various levels depending on the project. In some cases, industry participants have been involved in identifying the problem and objectives of the study. For those projects that require vessel time, industry involvement has been limited in some cases to the operation of the fishing vessel chartered to conduct the research. Recently, AFSC scientists have been able to establish two formal partnerships with nonprofit research foundations set up by industry groups. These partnerships involve the industry participants at all levels of the process from planning, execution, analysis, and reporting.

> “[In the northwest region] Cooperative research efforts aimed at developing new technologies to assess life histories for overfished species, improving bycatch reduction methods, assessing of fishing impacts on habitat, and generating more robust recreational data collection methods are other major needs.”

Impacts of cooperative research on science and management in the Alaska region

The ongoing effort with Alaska's trawl fisheries and the North Pacific Fishery Management Council to address the halibut bycatch in the various groundfish fisheries is an example of the impact of cooperative research on fisheries conservation in the Alaska region. Since 2002, this cooperative research partnership has also focused on factors affecting bycatch of Chinook salmon *Oncorhynchus tshawytscha* and chum salmon *O. keta* in the walleye pollock *Theragra chalcogramma* midwater trawl fishery. By understanding the avoidance behavior of salmon using underwater video and acoustic cameras, this partnership set out to design and test the effectiveness of various salmon excluder devices to reduce salmon bycatch without significantly lowering catch rates of pollock. Current excluder device designs have achieved a 40% reduction in Chinook salmon bycatch, while only lowering the pollock catch rate by 2%. The goal of this ongoing research is to achieve a 60–80% reduction of salmon bycatch, while keeping pollock losses to less than 5%.

Development and deployment of the salmon excluder device along with a voluntary rolling "hotspot" closure system is expected to result in a significant reduction in the Chinook salmon bycatch. Such a technological fix may obviate the need for existing savings area closures. This change in management would decrease the operational costs currently imposed under the existing regulatory regime on the pollock fleet and processors, while still meeting the council's intent to minimize the impacts to salmon in the eastern Bering Sea.

Future needs for cooperative research in the Alaska region

In Alaska, the NOAA Fisheries Service's goal is to establish multiyear projects with fishing groups representing most of the major gear/fishery components of the Gulf of Alaska and Bering Sea crab and groundfish fisheries. These projects will be implemented through the formation of additional nonprofit foundations via NOAA cooperative agreements or joint project agreements. In this way, industry and agency scientists can further pool their resources to conduct research projects as part of a long-range research plan. Plan development would be a cooperative process. A range of projects, including conducting species-specific resource surveys, collecting biological data to improve population parameter estimates for stock assessment modeling, collecting fishery data through electronic logbooks and observer programs, improving bycatch reduction technology and methods, evaluating the effects of fishing on essential fish habitat, and quantifying essential fish habitat would be some likely candidates under the plan.

Pacific Islands region—current program

Programmatic funding from the NCRP for this region is currently used to support the Northwestern Hawaiian Islands (NWHI) lobster-tagging project. This is an ongoing cooperative research project that uses chartered commercial fishing vessels as scientific platforms for the tagging and release of spiny lobsters *Panulirus* spp. and slipper lobsters *Scyllarides* spp. The scientific objectives of the tagging program are to (1) collect essential life history and distributional data for spiny lobsters and slipper lobsters, (2) provide requisite data to advance spatially explicit population dynamics model development, (3) document habitat degradation and associated bycatch (vertebrates and invertebrates) stemming from lobster fishing in the NWHI, and (4) supply marine vertebrate and invertebrate samples to elucidate trophic linkages and ecological dependencies. The use of chartered commercial fishing vessels is pivotal to the success of this program, as the annual research survey aboard the NOAA research vessel *Oscar Sette* lacks sufficient fishing effort to provide adequate tag recoveries for life history and abundance estimation (as well as samples for tagging).

> “In Alaska, the NOAA Fisheries Service's goal is to establish multiyear projects with fishing groups representing most of the major gear/fishery components of the Gulf of Alaska and Bering Sea crab and groundfish fisheries.”

Impacts of cooperative research on science and management in the Pacific Islands region

The lobster tagging program has provided data to reestimate life history parameters including natural mortality, growth, and movement. Previous estimates were dated. The program's analysis of growth parameters indicated a significant decrease over time at certain locations. These differences in time and space are necessary elements of an effective management program, especially when spatial management is required. Acquisition of tagging data has also allowed advancement of population model development resulting in more robust stock assessments. An integrated assessment model has been developed that specifically allows for the incorporation of tagging data, and this has resulted in less uncertainty when assessing stock status.

The cooperative lobster tagging project has used a bottom release device to safely return tagged lobsters to their benthic habitat and avoid incidental predation. Initially, fishermen indicated that they would be opposed to the use of this device in commercial operations for fear of a slow-down in normal operations and a concern that the benefits of its use were unsubstantiated. However, through first-hand experience (observed via video), fishermen now openly endorse the use of this device and may incorporate this device into their operations—should the fishery reopen.

> “[In the Pacific Islands region]...the future direction for cooperative research will be to establish a cooperative deep-slope resource survey to cover the entire Hawaiian archipelago.”

Future needs for cooperative research in the Pacific Islands region

In the Pacific Islands, the future direction for cooperative research will be to establish a cooperative deep-slope resource survey to cover the entire Hawaiian archipelago. This survey will collect abundance and biological data—primarily on crustaceans and bottomfish—as well as ecological and habitat data throughout the archipelago. Data deficiencies exist for many insular species listed in the FMPs, resulting in stock status determinations with large uncertainties. A further expansion of this survey to include the Samoan and Mariana archipelagos, as well other Pacific remote island areas under the jurisdiction of Western Pacific Fishery Management Council, is another long-term priority.

In summary, the NOAA Fisheries Service remains dedicated to ensuring that cooperative research is an integral part of a well-rounded marine science program. As we move into the era of ecosystem-based fisheries management, partnering with industry, academia, and nonprofit organizations will become more important in securing the necessary support for the successful stewardship of living marine resources.

American Fisheries Society Symposium 52:51–57, 2006

Session 1A: Making a Difference by Working Cooperatively

Management and Scientific Impacts, Panel Summary

Ralph Rayburn
Texas Sea Grant
2700 Earl Rudder Freeway South, #1800
College Station, Texas 77845, USA

> With the movement toward ecosystem based management, it will take a community to manage the fisheries. Through cooperative research and management there has been a fusion of three cultures: management, science, and fishing industries. [John Williamson, former member of the New England Fishery Management Council and member of the National Research Council committee on cooperative research]

The initial session of the joint Sea Grant/American Fisheries Society symposium focused on the management and scientific impacts of cooperative research. Ralph Rayburn, associate director and extension program leader for the Texas Sea Grant College Program and member of the secretary of commerce's Marine Fisheries Advisory Committee, moderated the session. Below are some highlighted comments from the oral presentations of the keynote and panelists, followed by a summary of the facilitated panel-audience discussion. Dr. Jim Balsiger framed the session for the panelists with a keynote address. White papers prepared by the keynote and panelists follow this panel summary.

- Cooperative fisheries research programs have existed in the National Marine Fisheries Service (NMFS) for most of its history. They are changing through time and have had many scientific impacts.
- The National Research Council panel studied cooperative research in NMFS in 2004. Recommendations included (1) the need to have expectations laid out with clearly defined goals, including what will happen with the data and results; (2) listing detailed responsibilities of parties involved; and (3) streamlining and standardizing all permitting procedures for conducting cooperative fisheries research.
- Cooperative fisheries research projects need to be collectively planned and executed so that there is increased buy-in and understanding on all sides. Working together is an efficient way to leverage limited resources and to use re sources most efficiently.
- Cooperative fisheries research must be integrated into the current scientific program and

Jim Balsiger, Alaska regional administrator and acting deputy assistant administrator for regulatory programs for the NOAA Fisheries/National Marine Fisheries Service (substituted for Dr. Bill Hogarth, deputy administrator for fisheries, NOAA Fisheries

needs to undergo a peer review process that is held to the same standards as all other science. The cooperative process needs to be documented relative to the design, development, execution, and integration into fisheries management. Projects must be conducted in a timely manner.

- Difficulty in utilizing cooperative research in fisheries models hinges on the need for long-term data sets; the current models that are used require longer-term data. We will get over this hurdle in 30 years, but it will be hard to input cooperative fisheries research data in the first few years; this does not mean that cooperative fisheries research data is being downgraded.

Summary and Overview of Panel-Audience Discussion

At the completion of the presentations, the moderator facilitated a discussion to address how fisheries are benefiting from cooperative research and management, what scientific and management impacts have arisen, and what new attitudes, knowledge, or hypotheses have emerged from collaborative programs.

- The Mississippi Interstate Cooperative Resource Association (MICRA, established 1989) was conceived by fish biologists who wanted to improve their performance in managing fisheries on an ecosystem scale, representing 28 states, two federal agencies, and two tribes. The mission is to conserve, develop, manage, and utilize interjurisdictional resources of the basin through true coordination and communication.
- MICRA works because there is recognition by all managing agencies that there is a need for cooperative fisheries research on an ecosystem scale. Other elements of MICRA's success are
    - state fish agencies control the agenda to prevent federal intrusion on their management,
    - keeping overhead low and leverage funds from many sources,
    - minimizing bureaucracy within the association to maintain the ability to adapt to change well and respond quickly by staying lean and efficient, and
    - the members are dedicated to the concept and mission of MICRA, cooperative fisheries research and management.

Mike Armstrong, director of Arkansas Fish and Game and chairman of the Mississippi Interstate Cooperative Resource Association

- Family fishing heritage traces two centuries to the coast of Croatia, with a history of fisheries innovation from hydraulic blocks to netting.
- In the 1990s, he recognized a change coming, so he and his son developed and received funding to design and test a variety of selectivity grid panels in the seine fishery. The most significant challenge was the limited time available to conduct testing and a large amount of gear.
- Need to do more research on size, material, and shape of grid panels, which is a challenge considering the number of fishing days that they are limited to in the fishery.

John Brajcich, owner and skipper of the FV *Franciscan I* of British Columbia, Canada

- Challenge to conduct bycatch species catch rate studies, quantify shrimp fishing effort estimates, and develop new gear technology to remove bycatch from trawls. Worked with Gulf and South Atlantic Fisheries Foundation for observer studies and gear testing along with the National Marine Fisheries Service and the Texas Shrimp Association.
- LGL Ecological Research Associates developed electronic logbooks as passive trip data collection devices because existing estimates of red snapper *Lutjanus campechanus* and shrimp industry bycatch were insufficient to rationalize differences.
- Through the cooperative fisheries re search program, an estimate of magnitude and age distribution of red snapper bycatch was developed that enabled a consensus between government and industry.
- Cooperative fisheries research had a valuable impact and was important in this case because it helped the industry realize that participation in such a program does not necessarily lead to more restrictive regulations.

Benny Gallaway, president of LGL Ecological Research Associates in Bryan, Texas

John Williamson, commercial fisherman, charter boat captain, former member of the New England Fishery Management Council and member of the National Academy of Science panel on Cooperative Research

- Conservation engineering is emerging from the fishing industry, which enables, through cooperative fisheries research, more fishing professionals to access and use the tools of science. In New England, there have been more than 200 cooperative fisheries research projects since 1999 in conjunction with NOAA Fisheries and the Northeast Consortium; this is more "collaborative" research than "cooperative" research.
- Acknowledgement that fisheries management efficiency will increase with adapt ability and that experimentation has facilitated greater fishermen involvement.
- As a result of cooperative fisheries research, data sets are broader, new and different ideas are tested, and the questions asked are of a different nature and much more interdisciplinary than have been in the past.
- With the movement towards ecosystem-based management, it will take a community to manage the fisheries. Through cooperative fisheries research and management, there has been a fusion of three cultures: management, scientific, and fishing industries. When fishermen are presented with scientific methods and tools, they are empowered and start to use the tools and partner with scientists.

Benefits

The session discussion focused on how the fisheries industry has benefited from cooperative fisheries research and management. John Brajcich said that he participates in cooperative fisheries research because he "wants to speak for the fish; fish are the basis of the fisherman's being and that is what motivates him. Cooperative fisheries research gives stakeholders the opportunity to work for the fish, fishery and the ecosystem." Mike Armstrong mentioned the long-range benefits to sharing information across political boundaries to include a broader and more interjurisdictional focus; working in this way spreads the resources, provides more money, and allows more questions to be answered.

Another benefit from cooperative fisheries research that was observed includes increased awareness through industry involvement in research. Benny Gallaway shared his experience in cooperative fisheries research where at first the fishermen did not believe the managers on such issues as sea turtle bycatch. Education and outreach efforts are important, but they were not all that was needed. Fishermen, at times, did not necessarily believe educational demonstrations. It was not until fishermen were involved in all parts of the science that they tended to have a greater belief and buy-in to the results. Another very important benefit of working cooperatively is that progress has been made where fishermen see that involvement in cooperative fisheries research does not necessarily lead to more restrictive management. The knowledge has gone both ways; fisheries

- Cooperative fisheries research has been a major focus for the Gulf and South Atlantic Fisheries Foundation as a means to consolidate what would otherwise be fragmented research efforts.
- Industry supported survey/observer efforts are good examples of how cooperative research has led to an improved assessment of bycatch levels in the Gulf and South Atlantic shrimp fishery.

Gary Graham, fisheries extension specialist for the Texas Sea Grant College Program and member of the National Academy of Science panel on cooperative research (substituted for Judy Jamison, executive director of the Gulf and South Atlantic Fisheries Foundation)

managers have observed that fishermen's perceptions on some critical issues are valid and must be considered in management decisions.

The regional fishery management councils are an example of one model for cooperative management. John Williamson noted that the New England Fishery Management Council is highly political and that there is a lot of industry involvement in management with very difficult, at times contentious, issues. Cooperation is not a new concept, and when people understand that management is necessary for the sustainability of the resource, they are more likely to accept regulations. One example Williamson expanded on was a cod-tagging program where two stocks were being ineffectively managed as separate entities. There was a need to identify the complex interactions between these two stocks. Through cooperative fisheries research, groups started to work together to develop and test hypotheses; through the cooperation of five institutions, a multi-tiered cod stock structure was elucidated and this allowed for a more appropriate management regime to be developed.

In summary, the discussion resulted in a list of benefits from cooperative fisheries research and management, including

- rewarding work that results in benefits sustaining the resource,
- greater information sharing,
- more interjurisdictional focus,
- the ability to address a greater number and broad spectrum of questions and issues,
- leverage resources,
- increased awareness of scientific method among industry,

- increased awareness of fishing practices and fishermen knowledge among scientists,
- greater acceptance of research findings, and
- more appropriate and effective management tools.

Impacts

Many examples were discussed where beneficial scientific impacts have resulted from cooperative research (CR). Excluder devices are now being supported proactively to increase protection of sea turtles, including funding the protection of Kemp's ridley nesting areas. This has improved the species abundance. Management impacts of CR were also discussed. Jim Balsiger talked about the different types of questions that are being asked and addressed because of increased cooperative research. He noted that the questions addressed through CR perhaps would not have otherwise been thought of, and stakeholder engagement has contributed to the process resulting in an expansion of research programs that inform management. Data sets are broader and expanding. This leads to a greater range of management alternatives, with potentially more robust data. John Williamson spoke about a project he is involved with called the Marine Resource Education Project, which was established 4 years ago to train fishermen in scientific methods, stock assessment models, and other management tools. This type of educational program allows fishermen to take a much more active role in the research process.

Ecosystem-Based Management and Cooperative Fisheries Research and Management

As fisheries management is moving from single-species-based management to ecosystem-based management (EBM), it is appropriate to consider the role of cooperative fisheries research and management. Mike Armstrong stated,

> At its core, there has been a paradigm shift in fisheries management. We do not manage on political boundaries, we now look at EBM, and so we need a broader scope. Cooperative fisheries research has guided thoughts in managing on ecosystem boundaries. Without the cooperative fisheries research, the twenty-eight states involved in MICRA have different, smaller questions, and for the tough questions, they tend to fall back into a parochial more fragmented view. We need to create structure and manage in an EBM way that stops fragmentation.

There are many lessons to learn through experience in cooperative fisheries research that may carry over and guide the design and implementation of EBM. John Williamson stated,

> Management is an experiment. As we move into EBM, there are no definite cause and effect relationships. We need to approach management in that realm, in an adaptive way. We need a data rich approach to set up a management action plan as part of a research plan and vice-versa. We need to take management action and set up monitoring efforts within the management plans to see if the efforts are effective. Cooperative fisheries research provides us with platforms, which involve people who have a stake in the resource. We need to monitor and make sure management action is working, and if it is not working, we should refine the regime. Throughout all of this, we should keep EBM in mind with all of its complexity.

Benny Gallaway stressed,

> When looking at EBM, it is important to keep in mind that the entire system would be completely different with different management regulations. Managing on the ecosystem based scale will be very difficult."

Jim Balsiger stated,

> Ecosystem-based management has been helped by cooperative fisheries research, because we engage different stakeholder groups in the process and the questions are much broader. You can't just look at

> uses of the natural marine resources in the ocean; you have to also look at the processing industry, the communities and the shipping industry. These things may not have been brought to the table without CR. It is a start. EBM is going to extend beyond the traditional way of management.

Challenges and Advice

The panelists discussed many examples of successes in cooperative fisheries research and management, but it is important to also recognize and learn from the less successful examples. Gary Graham shared an example where the shrimp fishermen's concept of the industry's impact on sea turtles was unsubstantiated and not realistic. While efforts were made with observers and various uses of turtle excluder devises to stimulate awareness of the fishermen's impact on turtle populations, this project failed because the study was not completed, and as a result, there are fishermen in various areas who feel the impact of the industry was isolated to only a few locations. He thought that if education and demonstrations were continued, awareness would have increased and opposition to regulations would decrease.

The partners involved largely influence the successes or failures of cooperative research and management programs. The relationships, trust, and commitment to programs are integral. Mike Armstrong noted that

> Social relationships are important in cooperative fisheries research. Management of partnerships is social management. In MICRA, they found more effectiveness in managing fisheries than in managing partnerships or relationships. Inclusion of social scientists is very important when evaluating cooperation.

Several panelists mentioned challenges that face cooperative programs. Specifically, funding, maintaining focus during decision-making processes, and vulnerability to political influences were observed to challenge programs. As more cooperative programs are being efficiently run, solutions are being slowly found to alleviate these challenges.

American Fisheries Society Symposium 52:59–63, 2006

Mississippi Interstate Cooperative Resource Association

MICHAEL L. ARMSTRONG
Arkansas Game and Fish Commission
No. 2 Natural Resource Drive, Little Rock, Arkansas 72205, USA

JERRY L. RASSMUSSEN
U.S. Fish and Wildlife Service
Post Office Box 774, Bettendorf, Iowa 52722, USA

The Mississippi River and its tributaries comprise one of the largest and most valuable ecosystems in the world. The basin encompasses more than 1.25 million square miles across all or portions of 32 states draining 41% of the continental United States. The basin extends from Montana to western Georgia within four major interjurisdictional subbasins: the Missouri River, the Arkansas River, the Tennessee River, and the Ohio River. Fish stocks in these waters provide important economic benefits to both recreational and commercial fishing.

The basin's interjurisdictional rivers and their floodplains are home to 102 species-recognized as threatened or endangered by the U.S. Fish and Wildlife Service (USFWS). The basin serves as a major pathway for the spread of aquatic invasive species such as the zebra mussel *Dreissena polymorpha* and Asian carp. Habitat alteration for navigation, flood control and power generation is widespread on the Mississippi River and each of the major interjurisdictional tributaries. Water management and flow regulation have been historically contentious and often litigious. Fish stocks moving between state, federal, tribal, and private management jurisdictions create complex resource management problems related to the management of important recreational and commercial fish stocks, protection of species of special concern, and prevention of the introduction and spread of aquatic invasive species.

The Mississippi Interstate Cooperative Resource Association (MICRA) was formed in 1989 by 28 of the basin's 32 state fishery management agencies to address interjurisdictional fishery management issues through a basin-wide or ecosystem management approach. The organization was conceptually conceived by the American Fisheries Society (AFS) Fisheries Administrators Section during discussions in the development of the Fisheries Action Agenda for the United States.

In addition to the 28 states, MICRA membership also includes three federal agencies (USFWS, U.S. Bureau of Reclamation, and U.S. Geological Survey/Biological Resources Discipline [BRD]), the Tennessee Valley Authority, and two Native American tribes (Chippewa-Cree in Oklahoma and Chickasaw Nation in Montana). The USFWS provides a part-time coordinator to handle administrative support, coordination, and grant management. MICRA activities are managed through an executive board with members representing each of the major subbasins entities currently operating within the basin: the Upper Mississippi River Conservation Committee, the Ohio River Fish Management Team, the Missouri River Natural Resources Committee, the Lower Mississippi River Conservation Committee, the Tennessee River Work Group, and the Arkansas-Red River Work Group. Although no formal working relationships exist between the subbasin entities and MICRA, it is this subbasin step-down structure that gives MICRA its unique ability to formulate policy and communicate issues across geographic and political boundaries.

“The Mississippi Interstate Cooperative Resource Association (MICRA) was formed in 1989 by 28 of the basin's 32 state fishery management agencies to address interjurisdictional fishery management issues through a basin-wide or ecosystem management approach.”

MICRA members developed a Comprehensive Strategic Plan and defined the organization's mission as to "Improve the conservation, development, management and utilization of interjurisdictional fishery resources in the Mississippi River Basin through improved coordination and communication among the responsible management entities." Interjurisdictional fisheries are defined as those species that depend on interjurisdictional rivers during some part of their life cycle and therefore come under the management of two or more government entities. In brief, MICRA's Comprehensive Strategic Plan (Rasmussen 1991) established the following 10 major goals:

1. Develop a formal framework and secure funding for basinwide networking and coordinating.
2. Identify and prioritize issues of concern for coordinated research that supports cooperative resource management (Rasmussen 1992).
3. Improve communication and coordination among entities responsible for fisheries resources.
4. Develop an information management program based on standardized methods.
5. Identify and coordinate fishery management programs to address species and habitat concerns from an ecosystem perspective.
6. Develop public information and education programs to disseminate information that supports fishery resource management.
7. Determine and document the socioeconomic value of fishery resources and related recreation.
8. Preserve, protect, and restore fishery habitats basinwide.
9. Develop compatible regulations and policies for fishery management to achieve interstate consensus on allocation of fishery resources.
10. Develop protocols, policies and regulations for disease control, introduction of exotics, maintenance of genetic integrity, and maintenance and enhancement of indigenous species.

MICRA's initial focus and galvanizing issue was to improve paddlefish *Polyodon spathula* management (MICRA Paddlefish/Sturgeon Committee 1993). In the late 1980s, the species had been petitioned for listing as a federal threatened and endangered species, and the lack of information on paddlefish stocks and lack of a coordinated effort to restore paddlefish within the basin were identified by both state and federal fishery managers as major challenges. Paddlefish stocks were believed to have declined dramatically throughout much of the basin due to habitat alteration and overfishing. Added fishing pressure on the remaining stocks was feared with collapse of the beluga sturgeon *Huso huso* caviar fishery in the Black Sea, a situation exacerbated by the disintegration of the Soviet Union. Paddlefish roe was increasingly being sought by industry as a preferred alternative to Russian caviar. Several states along the Ohio, Tennessee, and lower Mississippi rivers allowed the commercial harvest of paddlefish under a variety of regulations, while other states in the upper portions of the basin had closed all paddlefish harvest and some states were working to restore the species to portions of its former range where it had been extirpated.

"...Improve the conservation, development, management and utilization of interjurisdictional fishery resources in the Mississippi River Basin through improved coordination and communication among the responsible management entities."

MICRA initiated an interstate and interagency project to develop a basin wide paddlefish informational database in 1994 through the cooperation and financial assistance of the USFWS. The study actually commenced in March 1995, collecting and tagging adult paddlefish in the field and tagging any hatchery-released fish with a special binary, coded wire tag carrying a MICRA numbering system (Oven 1995). This mechanism was designed so that recaptures could be used to identify various stocks and to document movements, growth, mortality, harvest, and overall condition of the various stocks (Heinricher-Oven and Fiss 1996). By late 1997, a total of 22 states and several universities were participating in the study, and more than 6,000 wild, adult paddlefish had been tagged and released in the field, while more than 1 million hatchery-reared fingerlings had been tagged and released (Bettoli and Brennan 1997). Tag returns are increasing each year, and preliminary information indicates that these amazing fish are, as expected, highly migratory and are more effectively managed interjurisdictionally rather than on a state by state basis (Grady and Conover 1998). MICRA's Paddlefish/Sturgeon Committee continues to work on this project and has now initiated efforts to assess shovelnose sturgeon *Scaphirhynchus platorynchus* stocks within the basin (Hesse and Carreiro 1997). The information and communication among committee members has led to a refinement of paddlefish management strategies among the member states.

Aquatic invasive species are a major threat to the natural resources of the Mississippi River basin. MICRA member states formed an Aquatic Nuisance Species Committee and identified 46 invasive species as causing, or having the potential to cause, significant detrimental impact to native aquatic fauna. The problems incurred from zebra mussels alone in disrupting nutrient cycling within infested waters and increasing the economic costs of water use have been well publicized. MICRA has successfully elevated the issue of Asian carp (bighead carp *Hypophthalmichthys nobilis*, silver carp *H. molitrix*, grass carp *Ctenopharyngodon idella*, black carp *Mylopharyngodon piceus*) to a national dialog and has been a major player in the current effort to develop a national management plan for Asian carp. MICRA officially requested the listing of black carp as an injurious species under the Lacey Act in 2000 and has supported similar listings for silver carp and bighead carp. In 2001, the National Aquatic Nuisance Species Task Force asked MICRA to facilitate establishment of the Mississippi River Basin Panel of the Aquatic Nuisance Species (MRBP). MICRA continues to provide administrative support for the MRBP and disseminates information on the panel through the MICRA Web site.

"MICRA member states formed an Aquatic Nuisance Species Committee and identified 46 invasive species as causing, or having the potential to cause, significant detrimental impact to native aquatic fauna."

MICRA has achieved numerous other significant measure of success in establishing itself, in developing communication lines between members, in initiating field projects, and in establishing positions on issues of basin-wide concern (Brouha 1996; Rasmussen 1999). The following summarizes some of those accomplishments:

- A newsletter (*River Crossings*) is published six times per year with distribution nationally and internationally in both hard and electronic copy.
- A Web page is maintained with postings of MICRA position statements, reports, and activities.
- Committees have been established to address basin-wide paddlefish/sturgeon, aquatic nuisance species, freshwater mussels, and game fish issues.
- The Paddlefish/Sturgeon Committee developed a strategic plan for assessing the status of paddlefish and sturgeon in the Mississippi River basin.
- MICRA, through its Freshwater Mussel Committee, facilitated formation of the Freshwater Mussel Conservation Society and completion of the National Strategy for the Conservation of Freshwater Mussels. MICRA played a key role in elevating illegal shell poaching as a basin-wide issue and facilitated the formulation of standardized minimum length limit for certain mussel species in states with commercial mussel harvest.
- Grants totaling in excess of $350,000 from the International Association of Fish and Wildlife Agencies, National Fish and Wildlife Foundation, USFWS, BRD, and AFS have been used to fund a variety of projects, including paddlefish/sturgeon stock assessment, a mussel refugia study on the upper Mississippi River, an economic evaluation of navigation on the Missouri River, a catfish workshop and several nongame fishery assessment projects on the Missouri River.

- As a result of MICRA's Catfish 2000 Workshop (MICRA 1998) and follow-up projects of its Gamefish Committee, several of the states now promote and manage trophy catfish fisheries.
- MICRA has focused on improving riverine and floodplain habitats through participation on major national and international floodplain management panels and symposia, through information transfer on our Web page and in our newsletter and by providing close participation and support for the current National Fish Habitat Initiative.
- MICRA was invited and has become a member of the Midwest Association of Fish and Wildlife.
- MICRA has worked with the International Association of Fish and Wildlife Agencies to draft national legislation on species screening and to develop a national policy on the introduction of new aquatic species into the United States.

Lessons Learned

New knowledge

1. Increased knowledge of paddlefish stocks, and harvest status within the basin.
2. Knowledge of paddlefish biology, including movements over long distances within the basin, between subbasins, and out of Texas coastal tributaries into the Gulf of Mexico and back into the same or different tributaries.
3. The importance of managing paddlefish from a basin-wide perspective rather than limiting management concerns and actions within a single tributary of subbasin.
4. Increased awareness and concern with regard to paddlefish and sturgeon species and their exploitation by the world caviar markets.
5. Increased awareness of the ecological, economic and political challenges associated with the control and management of exotic species, particularly Asian carp, in the Mississippi River basin.
6. Increased awareness and concern for the basin-wide overharvest of mussel species.

Benefits

1. Improved communication between state and federal agencies concerning common issues: paddlefish, sturgeon, habitat, mussels, exotic species and game fish management.
2. Recognition and agreement that the resource management problems of the Mississippi River basin are more complex and larger than any single state or federal fish and wildlife management agency can handle alone and that the states and federal agencies can use MICRA to work together and to capitalize on the concept that there is strength in numbers and in the unity among partners.
3. Ability to use MICRA to elevate issues of concern among members to better communicate and leverage influence of other members to solve complex resource management issues.
4. Ability to operate more efficiently than government by quickly responding to the needs of its members and carrying projects and funding over more than one fiscal year.

Problems encountered

1. *Funding*—A major limitation to MICRA has been a consistent source of funding and the reliance on voluntary participation in projects. A secure and consistent funding source is needed to operate effectively for the long term. MICRA is currently exploring the option of seeking congressional authorization as a fishery commission.
2. *Bias for local focus*—MICRA's greatest challenge to date has been and will probably always be in dealing with problems related to the diverse resource management perspectives of its members. As a large interagency, interjurisdictional organization, MICRA needs to constantly be aware of the human tendency to think locally rather than globally.

3. *Decision making*—MICRA strives to operate under consensus, but resolving the differences between 28 member states and several entities is a complex procedure.
4. *Basin-wide versus subbasin issues*—MICRA constantly has to strive to focus on the core or common values and functions of most states within the basin, while being faced with myriad subbasin issues which can shift this focus off course.
5. *Political influences*—Vulnerability of MICRA members to political fallout resulting from MICRA positions on controversial issues is a constant concern.
6. *Dependence*—MICRA has had to rely on the services of a USFWS paid coordinator, placing it at the mercy of the federal administration and budgeting system.

References

Bettoli, P. W., and N. P. Brennan. 1997. Interim report on paddlefish coded wire tagging project April 1, 1996-June 1, 1997. Tennessee Technological Report submitted to Mississippi Interstate Cooperative Resource Association, Bettendorf, Iowa.

Brouha, P. 1996. MICRA - just do it! Fisheries 21(7):4.

Grady, J. M., and G. A. Conover. 1998. Mississippi River basin paddlefish research coded-wire tagging project: 1997 Annual Report. Mississippi Interstate Cooperative Resource Association, Bettendorf, Iowa.

Heinricher-Oven, J., and F. C. Fiss. 1996. MICRA national paddlefish research:1995 Interim Report. Mississippi Interstate Cooperative Resource Association, Bettendorf, Iowa.

Hesse, L. W., and J. R. Carreiro. 1997. The status of paddlefish, pallid sturgeon, lake sturgeon, and shovelnose sturgeon. Report prepared for the Mississippi Interstate Cooperative Resource Association, Bettendorf, Iowa.

MICRA Paddlefish/Sturgeon Committee. 1993. MICRA Paddlefish/Sturgeon Committee Strategic Plan. Mississippi Interstate Cooperative Resource Association, Bettendorf, Iowa.

MICRA (Mississippi Interstate Cooperative Resource Association). 1998. Catfish 2000, workshop proceedings. Mississippi Interstate Cooperative Resource Association, Bettendorf, Iowa.

Oven, J. H. 1995. MICRA - Mississippi Interstate Cooperative Resource Association - interjurisdictional paddlefish tagging study protocol. Mississippi Interstate Cooperative Resource Association, Bettendorf, Iowa.

Rasmussen, J. L. 1991. Interjurisdictional fisheries initiative for the Mississippi River Drainage basin - Comprehensive Strategic Plan. Mississippi Interstate Cooperative Resource Association, Bettendorf, Iowa.

Rasmussen, J. L. 1992. Mississippi Interstate Cooperative Resource Agreement (MICRA) activity prioritization - final report. Mississippi Interstate Cooperative Resource Association, Columbia, Missouri.

Rasmussen, J. L. 1999. What is MICRA and why does it work? Pages 218–229 *in* D. F. Williamson, G. W. Benz, and C. M. Hoover, editors. Symposium on the Harvest, Trade and Conservation of North American Paddlefish and Sturgeon. TRAFFIC, World Wildlife Fund, Washington, D.C.

American Fisheries Society Symposium 52:65–70, 2006

A Seiner's Experience Collaborating with Fisheries and Oceans Canada, Commercial Fishermen, and First Nations

John Brajcich
FV Franciscan 1
6295 Aubrey Street, Burnaby, British Columbia, Canada

Introduction

My family has a long tradition of seine fishing commencing with seining for sardella off the Dalmation Coast of Croatia in the 1800s. In 1924, my father emigrated to Canada and began seining for salmon and herring in British Columbia. In order to preserve this livelihood, he had to adapt to this new environment and respond to changes in the industry. This involved experimentation and taking risks. For example, in the 1940s, he built a new seiner, the *Freeland*, using a converted caterpillar tractor diesel engine with a new hydraulic system to power the winches. These mechanical additions were the first of their kind in the industry. In the 1950s, my family was the first family in Canada to use another new device, the Puratic Power Block. It, too, was hydraulically driven, and many fishermen believed it would not work; however, today it has become standard gear on all seine boats. My father continued to make modifications experimenting with new nylon netting and lines among other things. My brother and I followed in this tradition and, in the 1960s, built a new steel seiner, the *Provider*, which had a completely enclosed wheelhouse. At the time, many fishermen thought this was a foolish idea; however, it was so successful that most seiners built after the *Provider* had enclosed wheel houses and many existing boats were modified to include them. In the 1970s and 1980s, my brother and I built two more steel seiners, the *Dual Venture* and the *Franciscan no. 1*. These boats were outfitted with state-of-the-art electronics, and the fish hold was divided into four insulated compartments with a slush ice system for fish storage. These examples show a long family tradition of innovation in fishing. Perhaps all fishermen have to be innovative in order to survive, but current times require an innovation that is different from past innovations. Rather than increasing the efficiency in finding, catching and storing the fish, we now have to be innovative in how we ensure the sustainability of the resource, and this cannot be done alone. It requires working in collaboration with industry, managers of the resource, and scientists.

"Perhaps all fishermen have to be innovative in order to survive, but current times require an innovation that is different from past innovations. Rather than increasing the efficiency in finding, catching, and storing the fish, we now have to be innovative in how we ensure the sustainability of the resource, and this cannot be done alone. It requires working in collaboration with industry, managers of the resource, and scientists."

Change in the Fishery

In the 1990s, there was growing concern for certain coho salmon *Oncorhynchus kisutch*, Chinook salmon *O. tshawytscha*, and steelhead *O. mykiss* stocks in British Columbia. By 1998, there was scientific evidence suggesting serious declines in the Thompson River and the upper Skeena River coho salmon, and this "crisis" threatened to shut down the Pacific salmon fishery and forced a shift in paradigm that stimulated a greater focus on conservation than was seen in previous years. In response to this crisis, Fisheries and Oceans Canada (DFO) developed a strategy "to harvest available abundances of large, healthy stocks of salmon of all species while ensuring conservation of smaller, threatened stocks" (http://www-comm.pac.dfo-mpo.gc.ca/publications/selectivep_e.pdf). They implemented a more risk averse, conservation-based management strategy and outlined new selective fishing policies and a framework for implementation (http://www-comm.pac.dfo-mpo.gc.ca/publications/selectivep_e.pdf). The policy requires "licensed fishers to fish selectively, avoiding or releasing unharmed non-target fish, invertebrates and marine mammals." It requires that fishermen avoid critical areas such as nursery grounds or coho salmon interception areas. There are also special management zones where coho salmon nonretention is mandatory. According to this policy, if seiners are to continue fishing, there must be significant change to the current methods.

Selectivity grids in seine fishing

The foundation of the idea for using selectivity grids was laid in 1950 during a summer spent on my father's seine boat. One night, after all the marketable salmon was unloaded, I was in the fish hold discarding non-target fish. I counted 851 immature coho and Chinook salmon. All the other seiners unloading in the bay also faced the same problem. I remember thinking how wasteful this was, and this realization stayed with me throughout my career.

In 1998, during a discussion with my son, Paul, (a third-generation seine fisherman) about the new challenges facing fishermen, I asked him if it was acceptable for a logger to clear cut a hillside of trees. He replied that it was not acceptable. I then asked, "if it is not acceptable for loggers to clear cut hillsides why is it acceptable for fisherman to 'clear cut' salmon or other fish?" This comparison directed the discussion towards developing the concept of selectivity grids. By lacing grid panels into a seine bunt, the grids would allow the safe, in-water escapement of immature and non-target species while retaining target and marketable fish.

The Selective Fishing Program at Fisheries and Oceans Canada provided a venue where we could take this idea and test it. It encouraged stakeholders to be innovative and to share their experience and knowledge and thus created an exciting opportunity for us to be part of the solution by working in collaboration with DFO, other commercial fishermen, and First Nations. Initially, this program was funded through the Pacific Fisheries Adjustment and Restructuring Program, which had a budget of 21.5 million dollars over a 4-year period in addition to funding from 5% of the total allowable catch (TAC). This budgeted contribution wrapped up in 2002, and the work now continues being funded entirely by the 5% of TAC.

"The Selective Fishing Program at Fisheries and Oceans Canada provided a venue where we could take this idea and test it. It encouraged stakeholders to be innovative and to share their experience and knowledge and thus created an exciting opportunity for us to be part of the solution by working in collaboration with DFO, other commercial fishermen, and First Nations."

In 1999, Paul and I approached managers of the Selective Fisheries Program with a proposal to test the selectivity grids. The proposal included a budget and a plan to work in collaboration with managers of the resource, other commercial fishermen, and First Nations. The proposal was accepted and has been renewed and remodified every year, enabling us to build on the knowledge gained during each set of experiments and to make adaptations as our knowledge and experience grew.

Paul and I led the project and bore all the costs of the experiments, including those pertaining to logistical arrangements, supplies (such as netting, spreader bars, and grids), salaries for crew, training costs, chartering of boats, manufacturing, and data analysis. Our costs were then recovered through TAC.

Methodologies and Results

In the first experiment, there was one vessel that fished with four grids alongside a control vessel that fished with conventional gear. The boat with the grids outfished the boat without grids. The results suggested that fish were escaping through the grids. The next year, both vessels used grids, and then in subsequent years, there were five vessels (including First Nations participation) fishing in various locations using the same grids and testing opening size, opening shape, color and material, and location in the bunt. Results continued to suggest that the grids were an effective way of releasing the non-target fish (see below for more detail). The purpose of the additional vessels was to increase the sample size and introduce the new selective gear and fishing techniques to other fishermen seeking their feedback.

Grids

The material and construction of the grid was modified over time. It had to be strong enough to withstand the weight of large sets and flexible enough to address the demands of drum, spooling pin, and power block, while minimizing damage to fish. The awkwardness and extra care required while working with more rigid grids has shown that the soft, flexible, polyvinyl chloride (PVC) and Ethylene Propylene Dienterpolymer Membrane composites are the best design for seine fishermen.

Initially the grids were 24 × 20 in, and they have now been modified to 36 × 20 in. Preliminary testing has also been done on 72-in-long grids, which produced favorable results (see Figure 1.).

The color of the grid also made a difference. We discovered that there are six different colors of water on the coast of British Columbia and that the colors of the grid-impacted escapement rates. We concluded that the clear PVC proved the most favorable color by 65%.

We also experimented with the placement of the grid within the bunt when the additional vessels were included in the experiments. Each day, grids were exchanged; the two grids located in the middle of the bunt were rotated with the two grids located in the front of the bunt. This rotation of grids was repeated each day until a sufficient statistical sample size was achieved. Once this number was obtained, the next set of grids to be tested entered into the daily rotation. The additional vessels also

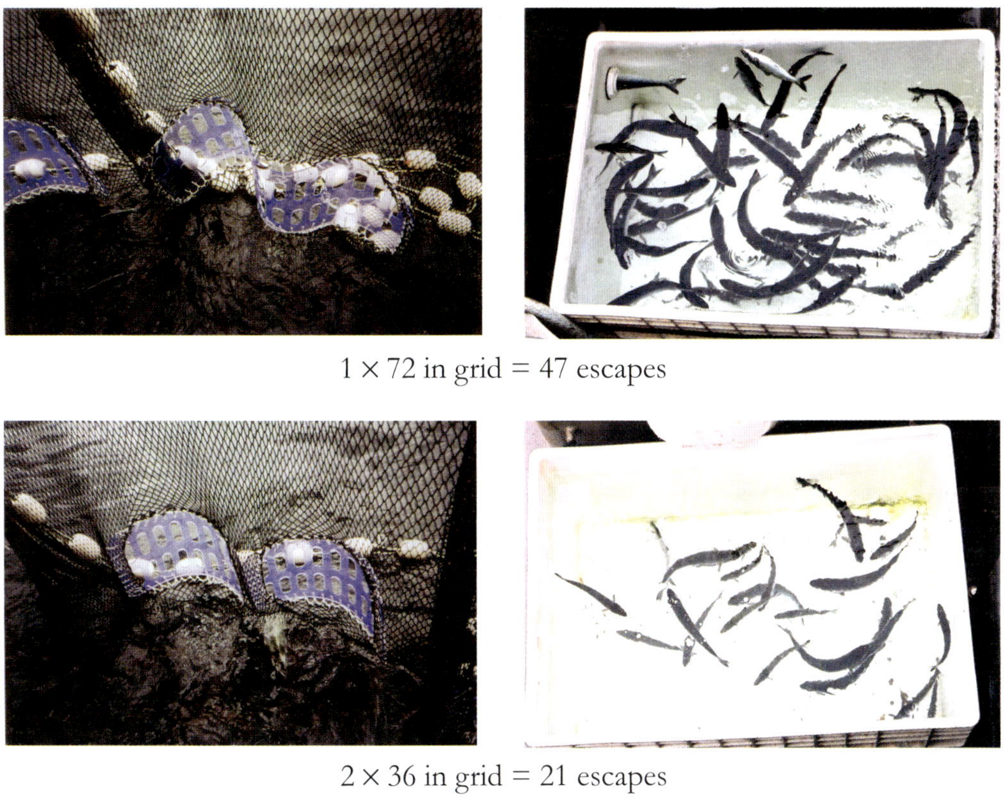

Figure 1. Comparison one 72-in-long grid with two 36-in-long grid and number of escapes.

revealed that vessel length affected the effectiveness of the grids, and the placement of the grids needed to be adjusted according to vessel length.

We experimented with different sizes and shapes of the openings and results show that the clear, flexible, 36-in-long PVC grids are the easiest to use and have produced the best results when compared to other sizes. The 36-in grids will be industry standard. Results also suggest that the oblong shape performs best to simultaneously release bycatch and retain target fish. However, further refining is recommended and will be reviewed once the technical report from the last round of experiments (2004) is complete.

Knotless bunts

We initially used knotted bunts. Knotless bunts were then introduced in 2002—these bunts were smoother and, therefore, less damaging to fish. Knotless bunts weigh less than knotted bunts but are not as strong and may be more difficult to unknot because they are woven. However, we found that knotless bunts eliminated loss of scales and loss of mucus to the non-target and immature fish. This was documented by observations of skippers who completed a questionnaire recording the loss of scales (e.g. skippers were required to circle one of the following observations describing the loss of scales: lots, some, very little).

Changing from ramping to brailing

Initially, we ramped the fish over the stern as traditionally done but then moved to brailing as we found this technique was less damaging to the fish. We also introduced brailing with water on sorting tables. The target species were placed into sorting boxes and the immature and non-target species were measured and then were resuscitated in revival boxes until ready for release. When the bunt was dried up, there was a hang time of 3 min per set, and then the grids were hauled out of the water. With the introduction of the brailing, measuring, and reviving, it required additional time to process than was required with ramping. For example, one experimental set took 3 h when there were 500 fish in the net to measure and sort.

Catcher bag

Catcher bags were introduced as a way to document the escapement. A catcher bag, made of very small mesh, was attached to the outside of the bunt, with separate chambers for the different grids. Catcher bags are used to contain all the fish that have escaped through the grid to determine the efficacy of the grid by documenting the size, species, and any preference to shape or design. Grids are size-selective, but with the addition of the catcher bag, the grids can also be species selective. For example, when targeting pink salmon *O. gorbuscha* that are running with chum salmon *O. keta*, the larger grid is used so that the chum remain in the bunt while the pink salmon escape to the catcher bags, reversing the normal process. In one instance, 51 small springs/Chinook salmon were caught and 48 went through grid into the catcher bag. In another instance, 720 pink salmon escaped through 48 holes in 3 min.

“It is our belief that there are three impediments to getting compliance from the entire seine fleet: lack of awareness and training, resistance to change, and financial hardship.”

What Remains to Be Done?

Peer review

Using the information obtained from these experiments, resource managers and scientists will formally write up the experiments in a scientific paper that will be peer-reviewed by the Pacific Scientific Advice Review Committee and then published.

Long-term mortality study

A preliminary long-term (48-h) mortality study has been conducted to determine if there are delayed mortality capture effects in the non-retention fish. This study involved taking fish that escaped through the grid and holding them in big pens, tagging them, and then releasing them after 48 h. These tagged fish were later captured in the United States and Canada sports fisheries, in river fisheries, and on the spawning grounds. The mortality rate for the fish escaping through grids was 10–25%, considerably lower than compared with a 70–80% mortality rate of fish caught and released from on deck during a commercial seine fishery. The next planned experiment will be with knotless bunts. An in-depth, long-term study on delayed mortality is also needed.

Getting compliance

It is our belief that there are three impediments to getting compliance from the entire seine fleet: lack of awareness and training, resistance to change, and financial hardship. Seine fishermen need to be informed about the success of the selectivity grids and trained in how to effectively use them. One way to address this could be the production of an information/training video or DVD that describes the concept and provides step-by-step training.

> "A small group of leaders in the industry have emerged and are setting an example of how to work in collaboration to find solutions to meet conservation needs and thus moving towards true cooperative management."

Many seine fishermen are also reluctant to change. This is in part due to a focus on short-term gain or an immediate return on their investment rather than a focus on long-term gain or sustainability. This might be overcome if there were incentives for compliance. For example, additional fishing opportunity (longer periods of time and availability in areas in which seine fishermen were previously excluded, e.g., nursery and/or rearing areas) for those using the gear could be an incentive to comply.

Some fishermen have also experienced decreases in income making it more difficult to purchase new gear and train crew. A way to address this could be to apply the license fee towards gear modification. Another way could be for DFO to supply the entire fleet with knotless bunts and one set of 36-in grids at an estimated cost of 1.25 to 1.5 million dollars.

Finally, if selectivity grids were a legal requirement, there would be complete compliance. To date, it has been difficult to have the use of selectivity grids mandated into policy because the current system requires that elected representatives of the fleet vote on whether to accept or reject recommendations for conservation measures.

Summary Conclusions

A conservation crisis stimulated the need to change the fishery. In an attempt to address this crisis, it was crucial for the industry members to work in partnership with resource managers, scientists, First Nations, and other industry. A small group of leaders in the industry have emerged and are setting an example of how to work in collaboration to find solutions to meet conservation needs and thus moving towards true cooperative management. Selectivity grids have proven to safely and successfully release salmon without related handling stress and are an outcome of this partnership. The in-water sorting system for seine fishermen clearly reduces and/or avoids encounters with immature and non-target species. This year, four boats are scheduled to fish in the open fishery with special permit to study the durability of grids in the normal fishing season. This use of grids has other applications as well. For example, Scripps Institute at the University of San Diego in California requested information about this project, and we provided them with a set of grids that could be used in the tuna seine fishery. Education and training, the continuances of the Selective Fishing Program (funding and/or funds from TAC) and incentives for compliance are needed to build upon the momentum achieved on this project and to further address the conservation crisis.

American Fisheries Society Symposium 52:71–73, 2006

Making a Difference from Working Cooperatively

Benny J. Gallaway
LGL Ecological Research Associates, Inc.
1410 Cavitt Street, Bryan, Texas 77801, USA

How Are Fisheries Benefiting from Cooperative Research and Management? In Your Opinion, What Impacts Have Cooperative Research and Management Had on Fisheries that Would Not Have Happened Without It?

My experience comes from the Gulf of Mexico where a penaeid shrimp fishery operates. This fishery typically ranks at or near the top of the nation's most valuable fisheries. The shrimp stock is healthy and not overfished. However, this bottom-otter-trawl fishery is plagued with bycatch problems. The bycatch includes endangered sea turtles and other economically valuable species, notably juvenile red snapper *Lutjanus campechanus*. Attempts to reduce shrimp-trawl bycatch to the extent practicable have included the development of turtle excluder devices (TEDs) and bycatch reduction devices (BRDs, pronounced "birds").

The introduction of TEDs in the late 1980s was a biopolitical nightmare that was extremely contentious and marred by civil disobedience and even violence. Fishers believed that turtle catches were too low to justify the shrimp losses associated TEDs and that TEDs were certainly not needed at all times and places. Ultimately, TEDs were required at all times and places fishing was conducted, except under limited circumstances.

Today, a much better process has evolved, namely cooperative research programs (CRPs). The shrimp industry, for example, has led recovery efforts for the Kemp's ridley sea turtle by directly supporting nesting beach protection programs with capital and in-kind support. This program being conducted cooperatively with Mexico and the U.S. Fish and Wildlife Service has been a major contributor to an exponential increase in Kemp's ridley nesting and population recovery. At the same time, cooperative studies are being conducted with the National Marine Fisheries Service (NMFS) to confirm the level of shrimp loss associated with TEDs. These studies are not being conducted to develop an argument for removing this gear from required use—economics are given very little consideration in the management of endangered species. However, economics can be taken into account when considering other gear modifications to reduce bycatch that may increase the existing levels of shrimp loss. Considering these preexisting losses has led to acceptance by industry that impact assessments now incorporate a "fairness factor" that was previously missing from the assessments.

"Today, a much better process has evolved, namely cooperative research programs (CRPs). The shrimp industry, for example, has led recovery efforts for the Kemp's ridley sea turtle by directly supporting nesting beach protection programs with capital and in-kind support."

Through CRPs, fishers are directly participating in some of the critical research that governs their fishery. What they hear from managers they do not necessarily believe; even what they see, they may doubt. However, they are much more likely to accept the results from programs that they have been involved with from the design phase through collection and analysis of the data to the reporting phase. Further, they are seeing that the results of these projects are being incorporated directly into management plans. The fishery is benefiting from better management resulting from better information, and this information is credible to the fishers since they gathered much of the information themselves as part of CRPs. All concerned have benefited.

What Management Impacts Have Arisen from Cooperative Research and Management?

In the Gulf of Mexico, several key management impacts have resulted from CRPs. For example, defined levels of shrimp fishing effort over time and space is a critical component for bycatch estimation. This historical method for estimating this effort has long been known to be subject to error, but the assumption has been that these errors would be random rather than statistically biased. A CRP conducted under the auspices of the Gulf and South Atlantic Fisheries Foundation, Inc. resulted in a major finding. The effort estimates were, in fact, statistically biased. Nearshore and shelf-edge effort has been systematically underestimated, and mid-shelf effort has been consistently overestimated. This finding led to a commitment on the part of the managers to obtain better data. As a result, a major CRP to directly measure shrimp fishing effort has been funded by Congress and implemented through the NMFS in conjunction with industry.

“Perhaps the most significant attitude change within industry is dispelling the belief that cooperating in scientific studies always results in an outcome that will lead to more restrictive management policies.”

Effort is only one side of the bycatch estimation equation; one also needs to know the catch rate of the bycatch species. NMFS conducts an observer program, and a complementary observer program is being conducted by industry. The collective results have yielded data enabling more credible catch rates for bycatch species and better definition of the age composition of the bycatch. Key findings from these programs have shown that age-0 red snapper (for example) dominate the bycatch rather than age-1 fish, age-0 fish enter the fishery earlier than suspected (formerly bycatch during the early season were counted as being age-1 fish), and natural mortality rates of age-0 and age-1 fish have been found to be approximately twice as high as formerly estimated. These results have led to the observation that while bycatch is an important factor contributing to the overfished status of red snapper, the directed fishery harvest is substantially more important than had been previously estimated. The management change is that directed harvests and bycatch must both be taken into account and that bycatch management alone is not sufficient to recover the stock.

What Scientific Impacts Have Arisen from Cooperative Research and Management?

The aforementioned changes in bycatch estimates, age composition of the bycatch, and natural mortality of the key bycatch species have resulted in expanding the scope of plausible factors that might explain the observed dynamics of species taken as bycatch. Simply put, a better understanding of the biology of the key species is emerging, and this will undoubtedly lead to better and more effective management. For example, papers published in the peer-reviewed literature by scientists representing industry in CRPs predicted that BRDs would be far less effective in reducing bycatch than was being estimated. This led to additional CRPs that have demonstrated that this is indeed the case. While the issue now is to develop a modified technology or a different approach that will be effective, we are avoiding the pitfall of believing that an ineffective device is actually producing significant bycatch reduction. No one would benefit under that scenario, least of all the stock.

What New Attitudes, Knowledge, or Hypotheses Have Emerged from Collaboration among Fishermen, Scientists, Managers, and Communities?

Perhaps the most significant attitude change within industry is dispelling the belief that cooperating in scientific studies always results in an outcome that will lead to more restrictive management policies. In this instance, the conduct and reporting of the CRP studies have benefited the shrimp fishing community. Further, the results have also led to a willingness on the part of scientists and managers to question previous concepts regarding the factors governing the dynamics of the bycatch species. For example, understanding the role of density dependence and the stage at which compensatory mortality processes occur has now become a priority issue. Most importantly, a greater level of trust among all the parties appears to be emerging, and this bodes well for future management.

American Fisheries Society Symposium 52:75–77, 2006

Making a Difference from Working Cooperatively: Gulf & South Atlantic Fisheries Foundation, Inc.

Judy L. Jamison
Gulf & South Atlantic Fisheries Foundation, Inc., Lincoln Center, Suite 740
5401 West Kennedy Boulevard, Tampa, Florida 33609, USA

How Are Fisheries Benefiting from Cooperative Research and Management? In Your Opinion, What Impacts Have Cooperative Research and Management Had on Fisheries that Would Not Have Happened Without It?

Through the use of vessel observers in the Gulf of Mexico and the South Atlantic, the Gulf & South Atlantic Fisheries Foundation, Inc. (Foundation) has been able to provide an incredible amount of scientific data that has been used by industry and fishery managers. It is important to note that in-depth vessel observer programs were not in existence prior to the 1990s in our region. The foundation took great strides to pioneer cooperative observer programs and has been extremely successful in achieving a high degree of collaboration within our fisheries. Through these efforts, we have been able to achieve trust and close working relationships with members of the scientific and management communities as well as fishermen.

One of the first research programs that involved foundation observers involved catch characterization of shrimp trawl harvest. The foundation worked with scientists and fisheries managers to develop a scientific sampling protocol to be employed aboard commercial trawlers. Cooperators from the shrimp industry were recruited to take observers aboard their vessels. Through this work, the most comprehensive data base of trawl bycatch that was ever created was developed through our observer program. Through data that were collected from thousands of tows, it was determined that overall bycatch from southeast shrimp trawlers was not at the magnitude projected by many at the time. A commonly used example was that 10 lbs of bycatch resulted from each pound of shrimp landed. Depending upon areas in the Southeast, the ratios were clearly shown to be closer to 3:1 or 4:1. This is not to say that bycatch is not a problem in the shrimp fishery, but our work did provide a more accurate assessment of trawl bycatch.

“The foundation has worked extensively with innovative shrimp fishermen to develop bycatch reduction devices that will reduce the take of unwanted finfish during shrimp operations.”

Our cooperative research efforts with turtle excluder devices (TEDs) highlight very significant work in the Gulf of Mexico and South Atlantic. Through cooperative efforts, more effective TEDs have been developed and demonstrated. The sea turtle issue has presented the most contentious problem ever encountered in southeast fisheries. Collaborative work with the shrimp industry has been of great benefit in addressing this problem. Tangible results from these efforts are being reflected in an exponential increase in nesting female Atlantic Kemps Ridley sea turtles *Lepidochelys kempii*.

What Management Impacts Have Arisen from Cooperative Research and Management?

The foundation has worked extensively with innovative shrimp fishermen to develop bycatch reduction devices that will reduce the take of unwanted finfish during shrimp operations. A major focus has been directed toward reducing the take of juvenile red snapper *Lutjanus campechanus* from trawl bycatch and has become a major challenge to our organization. Closely related to this problem is the impact of trawling to the red snapper resource. Through collaboration with the shrimp industry, an extensive data base regarding age structure and quantity of red snapper harvested in shrimp trawls has been developed through our observer program. These data have been important to fishery managers in determining the extent of impact from trawling and have been useful for determining fishery management measures in our gulf fisheries.

Another cooperative research project sponsored by the foundation focused on red snapper discards and related mortality from recreational head boats. This project provided insight to managers during a pilot effort directed to increasing the retention size of red snapper in the recreational fishery in order to extend the season in the western Gulf of Mexico. Through collaboration with the scientific community and head boat operators, onboard observers collected data on discards of undersized fish harvested on head boats. High discards associated with unacceptable release mortalities were documented, and this convinced managers to discontinue this management scheme.

"Perhaps the largest contribution to scientific impacts that we have provided relates to trust that the industry has for our data through cooperative research."

An important project supported by the foundation relates to a pilot project to monitor shrimp effort. Congress appropriated special funding to assist with this pressing issue. One of this symposium's panelists, Dr. Benny Gallaway, developed a pilot project utilizing electronic log books aboard trawlers to determine effort. Since completion of this project, the Gulf of Mexico Fisheries Management Council is considering implementing this tool, which was evaluated aboard a number of cooperating commercial vessels.

What Scientific Impacts Have Arisen from Cooperative Research and Management?

Some of the initial investigations that utilized vessel observers in the shark fishery of the Gulf of Mexico and South Atlantic were performed by the foundation. Fundamental data regarding catch per unit effort, catch characterization, and important biological parameters associated with sharks and its fishery were provided through our organization. Important scientific information was obtained through this, while at the same time, fishermen trusted the data that were collected from an industry organization and through a project in which they had substantial input in design and execution.

Perhaps the largest contribution to scientific impacts that we have provided relates to trust that the industry has for our data through cooperative research. Although data collected by the foundation is shared with the scientific and management communities, we employ our own data analyst. This independent industry approach provides fishermen, who often do not trust scientific interpretation, with confidence in our conclusions from scientific investigation.

What New Attitudes, Knowledge, or Hypotheses Have Emerged from Collaboration among Fishermen, Scientists, Managers, and Communities?

Another example of working cooperatively toward scientific and management goals relates to a highly emotional topic in our regional fisheries—endangered sea turtles. Tremendous debate raged in the western gulf regarding the interaction of trawls with sea turtles. The foundation initiated a project to investigate spatial and temporal distribution, as well as abundance, of sea turtles encountered in shrimp trawls. The shrimp industry provided significant cooperation in this study by allowing observers aboard their vessels to collect data from their trawls that had turtle excluder devices removed. Collaborating industry members were regulated to restrictive tow times so as not to impact harvested sea turtles. These short tow times did cost fishermen some production of shrimp. After a year-long study, sea turtles were documented as relatively abundant on the nearshore shrimp grounds and the use of turtle excluder devices highly warranted. Sea turtle interactions were also recorded in the deeper waters of the shrimp grounds, but not nearly as extensive in numbers as those in shallower areas. In addition to providing important data to scientists, results of these investigations assisted the industry in becoming more aware of the potential problems associated with sea turtle/trawl interactions. Although some industry members did not like the outcome of this project, they accepted the project results as accurate. The fact that this work was done in cooperation with fishermen and through an organization that they would trust served substantially to address concerns regarding this major problem.

I have discussed only a few of the numerous collaborative projects undertaken by the Gulf and South Atlantic Fisheries Foundation. We have learned that the success of cooperative fisheries research and management efforts hinges upon several fundamental, but highly important issues. First, it is very important that industry be included at the beginning of a proposed undertaking. Our Foundation Board of Trustees, composed of members from the fishing and seafood industry, serves as an excellent sounding board for proposed projects. Often input is requested from both the scientific community and fishing industry during early planning stages of collaborative research projects, and this is done through informal meetings. Early concerns expressed by either party are discussed, and every attempt is made to address any perceived problems. The foundation utilizes experienced field coordinators that regularly maintain contact with industry cooperators and the scientific community. These coordinators are in frequent contact with vessel captains and crew that participate in foundation efforts. Any problems discovered are rapidly handled. We have learned that communication and coordination are a very important component contributing to the success that we have experienced in the cooperative fisheries research arena. Research must be performed in a manner that regards scientific excellence. Throughout our projects, we communicate closely with members of the scientific community and seek advice and input regarding our ongoing cooperative projects. We have learned that an important element regarding collaboration relates to ownership of the project and its results. Industry must feel a degree of ownership regarding research in which it has participated. In order to accomplish this, the foundation regularly conducts workshops in which industry members participate. During these workshops, results of projects are discussed and the degree of their collaboration stressed. Additionally, we have learned that it is important for field coordinators to regularly confer with industry in the numerous ports located throughout the region. In many cases, the foundation supports regional efforts to disseminate research findings on a port-by-port basis through use of foundation coordinators. The key elements of planning, coordination, follow-up, and dissemination of results are thought to be imperative in foundation projects and should involve the scientific/management community as well as the fishing industry.

American Fisheries Society Symposium 52:79–82, 2006

Management and Scientific Impacts of Cooperative Research in the Northeast: A Cultural Fusion and Emergence of Cultural Translators

JOHN C. WILLIAMSON
Fishing Community Activist
201 Western Avenue. Kennebunk, Maine 04043, USA

How Are Fisheries Benefiting from Cooperative Research and Management? In Your Opinion, What Impacts Have Cooperative Research and Management Had on Fisheries that Would Not Have Happened Without It? What Management Impacts Have Arisen from Cooperative Research and Management? What Scientific Impacts Have Arisen from Cooperative Research and Management? What New Attitudes, Knowledge, or Hypotheses Have Emerged from Collaboration among Fishermen, Scientists, Managers, and Communities?

There has always been an element of cooperative research and management in the northeast; however, in the 1990s, we have seen an order of magnitude increase in collaborative research endeavors. As a direct effect, this has allowed greater reliance on cooperative management strategies as well. Therefore, we have seen two results from cooperative research and management that represent overarching change: an expansion of the entire data set for managing fisheries, and a fusion of cultures with the emergence of cultural translators in each of the three communities (science, management, and fishing).

"The cultural norms of fishermen and scientists are worlds apart."

Cultural Fusion

The cultural norms of fishermen and scientists are worlds apart. Both groups have equally valid viewpoints, but their different means of experiencing the marine environment often lead them to radically different perceptions. Assessment scientists view fisheries on a regional scale, while fishermen's experience is much more local. Scientists look at snapshots of data over a long time series, while fishermen observe a continuous-feed movie on a seasonal scale. Successful fishermen correctly intuit plans of action from multiple sources of information, while successful scientists systematically test hypotheses.

The fishing public has high expectations of science. In reality, these expectations may be misplaced, partially because of general misunderstanding of the scientific method. In the Marine Resource Education Project, John Boreman, director of the Northeast Fisheries Science Center, offers fishermen a seminar focusing on this miscommunication. Dr. Boreman's seven myths of science are as follows:

1. Science knows all.
2. Science is perfect.
3. Science advice is above reproach.
4. Models are crystal balls.
5. The more complex the model, the better.

6. Models always give us the right answer.
7. Science is too complex to explain.

John's seminar is a great example of the changing power relationship between scientists and fishermen (and scientists and managers). With the volume of collaborative research projects in New England (>200 since 1999) and the expanding numbers of fishermen and scientists entering into these partnerships, greater understanding of science concepts and methods are infused in the communities' approach to fisheries management problems. By encountering the limits to what science can do and tell us, it has had the effect not of undermining faith in science (which is what some people feared), but of directing the fishing community's attention toward those ways that fishermen can apply their knowledge to improve data. The scientific method and models are now recognized as accessible tools rather than unchallengeable abstractions.

The other aspect of cultural fusion is the emergence of translators, people who work at the interface between the science and fishing communities, or fishermen and the management process. In fishing communities, there are leaders and opinion-makers, respected on the pier or on the water, who are not so obvious to outsiders. Collaborative research is proving to be attractive to these local leaders, who are often early adopters by nature. In developing successful relationships with scientists and managers, they are disbursing the cloud of suspicion that the community has had of the science. Increasingly, we are seeing these people in the regulatory arena, and they are serving as a conduit of information on science and management developments.

In the Marine Resource Education Project, a fisherman facilitates each of two curriculum modules. During a 3-d curriculum module on fishery science, as an example, the fisherman/translator will have full control over the pace and agenda. He will have reviewed content in order to coach individual science presenters on their use of language/jargon. The translator's job is to draw the fishermen taking the course into intellectual parity with the scientists in the room. He will assist other fishermen in framing questions and will work with them to articulate fishery knowledge in the form of hypotheses for discussion.

Cod Tagging Program: A More Complex Stock Model

Similarly, scientists are being challenged to break out of old ways of thinking. Because fishermen's experience of the marine environment is of a different nature than the scientists', they are asking different kinds of questions, often challenging established assumptions. Many times, these have a direct link to management. An example of this is the cooperative cod-tagging program. Possibly the largest such effort ever attempted, more than 100,000 fish have been tagged over a 2-year period in a well-coordinated experimental design, using a wide range of vessels operating in specific locations and seasons. The impetus for this was a management strategy that defined separate stocks of Georges Bank and Gulf of Maine cod. Fishermen objected to what they felt to be an artificial and incorrect differentiation, questioning long-established stock boundaries based on growth rates and parasite loading. From a series of port meetings and interviews, a complex picture began to emerge. A collaborative task force was convened to determine questions of cod stock structure and how to test for them. The result of the tagging program will be a much more sophisticated stock model for the northeast and possibly more effective management.

> "...translators, people who work at the interface between the science and fishing communities, or fishermen and the management process."

Extraordinary Communication: Maine/New Hampshire Inshore Trawl Survey

Of course, the power of collaborative research rests in the vastly different skill-sets that fishermen and scientists bring to the job. The essential ingredient for ground-breaking collaborative research is that two disciplines communicate on an extraordinary level in order to accomplish things that neither could do alone. The Maine/New Hampshire Inshore Trawl survey offers an example.

Although NOAA Fisheries conducts trawl surveys of the East Coast exclusive economic zone, there is a need for more detailed information in nearshore waters. Until recently, no consistent long-term data set had ever been collected in state waters of Maine and New Hampshire; the logistics of sampling in the rugged coastal habitats of the Gulf of Maine had proven daunting. This represented a significant information gap affecting quotas, spawning closures, and other management measures.

The break through came from collaboration of state agency scientists and a team of experienced fishermen. Two identical 54-ft fishing vessels were made available for the inshore survey. An expert captain and crew were trained in survey methods. The vessels' owner and research captain, who were part of the survey design team, custom-built a trawl net and ground gear, designed to be efficient on short tows in a variety of terrain, for the range of species to be encountered.

In Maine, lobster traps dominate nearshore fishing grounds. What is more, the bottom is irregular, with limited patches of towable bottom interspersed with deposits of glacial till, boulders, and ledge. Protocols of a typical stratified random-station survey design do not work. In this case, a collaborative approach between the fishermen and scientists allowed them to invent an appropriate sampling procedure for the terrain. The sampling stations are randomly selected by computer from fixed strata well in advance of the survey cruise. These 100+ stations are plotted. From a nautical chart, the survey captain, a skilled fisherman, identifies the nearest likely survey tow for each randomly selected point. The supervising scientist and survey captain then visit communities in each survey area to consult with local fishermen. During interviews, the planned tows are adapted to knowledge of local conditions. Lobstermen are notified of the schedule so that they may remove gear from the tow paths. The Maine/New Hampshire Inshore Trawl Survey is in its fifth year.

> "...use of information generated in the collaborative research sphere needs to be deliberate and held to high standards....management planning must anticipate information needs well in advance and articulate these needs clearly and specifically to the fishing and research communities."

Deliberate Science for Planning

One focal point for collaborative research in the northeast has been the New England Fishery Management Council (NEFMC) Research Steering Committee (RSC), formed in 1999. A committee make-up unique for the council, the RSC is a collaborative effort of scientists from federal, state, and academic institutions, independent fishermen, council members with fishing backgrounds, and an environmental advocate. The RSC's task has been to shape an overarching management strategy for the NEFMC and to advise NOAA Fisheries on development of research programs to aid management. The culture of the RSC that has developed is striking. The fishermen members demand the most robust science as a prerequisite for management. The science members are just as insistent that the direction of research be responsive to community concerns and questions.

Beginning in 2000, the volume of collaborative research projects in the region began to build. Because project proposals are often in response to developments in fishery management, too often there is pressure to rush raw data into the decision-making arena. Therefore, the RSC and NEFMC have developed a procedure for managing the quality of these data. The RSC reviews final reports of collaborative research projects to ensure that project reports have undergone rigorous peer review; those that have not are directed to NOAA Fisheries to assign appropriate technical reviewers. As well, the RSC reviews the reports for quality of the work, lessons learned, and applicability to fishery management. These comments and the technical reviews become part of the project record. The RSC will then forward complete reports to appropriate management committees or forward recommendations for more extensive investigation to potential funding sources.

There is growing acknowledgement that use of information generated in the collaborative research sphere needs to be deliberate and held to high standards. This means that management planning must anticipate information needs well in advance and articulate those needs clearly and specifically to the fishing and research communities. It means that management actions should accommodate experimentation and incorporate monitoring and testing in an adaptive strategy. As information becomes available to managers, it needs to be reviewed for rigor, for its applicability to management, and for lessons to be applied to the next iteration of data acquisition.

American Fisheries Society Symposium 52:83–86, 2006

Making a Difference by Working Cooperatively: Social, Cultural, and Economic Impacts

PAT D. WHITE
Gulf of Maine Lobster Foundation
Post Office Box 523, Kennebunk, Maine 04043, USA

I was 16 when I bought my first lobster boat, built some traps, and started fishing off Marblehead, Massachusetts. That was in 1956. Since then, I have been involved in a few other occupations, but lobstering is my calling, as a harvester as well as an advocate for the industry.

I am currently the CEO of the Maine Lobstermen's Association, a commissioner on the Atlantic States Marine Fisheries Commission, a member of the editorial board of the National Fishermen, and an advisory committee member to the Northeast Consortium.

In recent years, I have had the unique privilege of serving as a commissioner on the Pew Oceans Commission (POC). Along with other members of the POC, I traveled to many regions of the United States and met with hundreds of people who were tied to fishing in one capacity or another. I heard their stories of working and living in and around the ocean and their concerns with the current state of the fisheries. I also listened to their vision of the future. Because the POC and the National Commission on Ocean Policy have now completed the tasks required of them, the chairs of the two commissions have formed the Joint Commission Task Force of which I am currently a member.

> "Fishing is a science, and the knowledge fishermen gain from their daily experiences added to the information they have from prior generations is every bit as important and valuable as knowledge gained from other sources."

When I was asked to consider the question, "What difference has cooperative research and cooperative management made for fishing and coastal communities?" I immediately thought of one of my first experiences in fisheries research.

Bob Steneck, Lobster, and Me

Some time around 1985, I met Dr. Bob Steneck, a biologist at the University of Maine. He was unique in a number of ways. I would take him and some of his graduate students out to dive sites in my boat, and they would count lobsters. He helped me understand why he was doing what he was doing and what it meant in the grand scheme of things. He also spent a lot of time learning what lobstermen did and why. We talked and shared knowledge: a biologist and a fisherman.

Today, as I work with fisheries managers, I try to stress the importance of anecdotal experiences. Fishing is a science, and the knowledge fishermen gain from their daily experiences added to the information they have from prior generations is every bit as important and valuable as knowledge gained from other sources. Our science is a very practical science, based on years of experience. We generate an inherent understanding of how things work by daily observation of the ocean and its resources.

As fishermen become involved in management, they gain a better understanding and tolerance of that process. It works in two ways—as managers work with engaged fishermen, they gain a better understanding and tolerance of the fishing profession. Overall, all sides are trying to understand each other.

Mutual Appreciation Needed to Conduct the Research and Management We Need Today

This is no small feat because fishermen do look at the world differently. They are incredibly independent, and historically, they have not had the best working relationship with management of any kind. There is considerable distrust.

There is a natural transition from science, and cooperative research in particular, to cooperative management. Once fishermen appreciate what science is and what scientists do, they will become better cooperative management partners. Likewise, once scientists appreciate what fishing is and what fishermen do, they will become better cooperative management partners.

The result will be better, more effective management measures. Cost effective research gives us far more capacity to cover the ocean than government research vessels could ever achieve. The United States spends less than 3% of its research dollars on ocean research. We can get a bigger bang for our buck through cooperative research.

This was a common theme I saw across the country—fishermen participating in research, trying to carve out a place in management, taking ownership of their resource, and managing responsibly.

Several examples stand out. In California, a salmon stamp generates funds for research. The salmon fishermen contribute from $65.00 to $165.00, depending on their landings. They are involved with how the money is spent, the majority of which goes to stock restoration, and habitat protection. The fishermen we talked with were very supportive of the project. They felt that part of the proceeds of every fish they caught went towards replacing it.

Similarly, in Maine, seed funding is provided by the lobster industry. Twenty dollars of every lobster license goes into a fund for lobster research and replenishment. A board of industry members decides on how it is going to be spent, and the research is done with scientists and industry members.

These industry initiatives are important. They build an appreciation within the industry for the need for effective management. They also build an appreciation among scientists and managers for effective working relationships with those that they seek to manage—the fishermen.

“The fishermen we talked with were very supportive of the project. They felt that part of the proceeds of every fish they caught went towards replacing it.”

Cooperative research can and does contribute vital information to aid in more effective management. For example, when industry began developing new regulations for lobster management, we soon realized that uniform regulations for the entire coast would not work. Regulations that work in Jonesport, Maine would not work in Long Island Sound, New York. We must tailor our management measures to regional ecosystems and communities. Cooperative research and management helps us accomplish this.

Professional Growth and Evolution as a Fisherman

I grew professionally as I, a lobsterman, became more involved in research projects, participated with more industry associations, and engaged with managers and scientists more. I served on Maine Sea Grant's Advisory Committee, then the Northeast Consortium's Advisory Committee. I have engaged in policy activities and research funding decisions.

For me, the most heartening development lately is the enthusiasm for cooperative research and management among the younger fishermen. A generational shift is occurring in New England. Younger fishermen are willing to become responsible research partners and resource harvesters with more ease and comfort. Cooperative research helps this professional growth occur.

Management decisions in the next 2–3 years in the Gulf of Maine will set the tone for the future of our fisheries. Big boat, small boat, individual transferable quotas, and consolidation—all will be decided in the very near future. We harvesters cannot afford to miss the boat when it comes to being partners in the decision-making process. The nature of our fishing industry and our fishing communities could be changed forever.

Broad Industry Participation

Cooperative research and management helps grow the next generation of fishermen. For example, it is bringing the gear sectors together, opening lines of communication that were not there before, giving fishermen and scientists the opportunity to talk and build a mutual appreciation for each other. Trawlers and lobstermen. Gillnetters and scallopers. Everyone.

One example is a cooperative, inshore trawl survey conducted by industry and Maine Department of Marine Resources scientists in the Gulf of Maine. It samples an area that the federal government survey vessels could never reach. It collects data that could not be obtained in any other way. To achieve this, industry-wide cooperation beyond those specific vessels that conduct the tows is needed. When the survey comes through, the entire industry must mobilize and move set gear out of its way. That is cooperation for a common purpose.

“If we are truly going to change the climate of fisheries decision making and management and find the management solutions needed to sustain the resource, the fishermen, the fishing communities, and the conservation organizations need to make the same transition that fishermen, scientists, and managers are trying to make.”

A second example is an initiative I have developed in the Gulf of Maine called the “Fishtank.” Its purpose is to encourage fishermen from all sectors who care enough about the future of the fishing industry to sit down and develop an ocean ethic focusing on preserving our industry without destroying each other.

I strongly believe these opportunities, which arose in part out of the relationships built through cooperative research and management, will improve our ability to understand the ocean and its ecosystem by asking the right research questions. It gives everyone a voice, and it will lead to the creation of a fleet composition and resource harvesting strategy that works best for the complex ecosystem dynamics throughout the Gulf of Maine.

Others Need to Play a Role in Cooperative Research

Unfortunately, we are missing an important stakeholder in cooperative research. The conservation organizations need to get more involved by talking with and listening to fishermen and scientists. If we are truly going to change the climate of fisheries decision making and management and find the management solutions needed to sustain the resource, the fishermen, the fishing communities, and the conservation organizations need to make the same transition that fishermen, scientists, and managers are trying to make.

Social, Cultural, and Economic Impacts

To conclude, let me say a few things about the panel’s discussion on social, cultural, and economic impacts of cooperative research and management.

Why do fishermen and scientists participate and how do cooperative research or management build trust?

While, for many, the initial motive to get involved was to prove those scientists and managers wrong, I think it changed quickly. We found out why scientists ask the questions they do and they found out why we ask the questions we do. For many, particularly the younger generation of fishermen and scientists, cooperation is becoming the way you do things. I see younger scientists now trying to understand what fishermen are thinking about. I see the younger fishermen trying to understand the science of the industry because it is good for business, it is necessary in the competitive global business world of fishing, and it is good for the health of the resource. They are genuinely interested in what the other has to say; they are speaking a language they both can understand. They are creating a bridge, a different way of looking at the world, between science and industry.

Fishermen may never fully understand how the stock assessment models work, but if they can understand the information that goes into the model and why, it legitimizes the model. Now fishermen, through cooperative research and management, can participate in collecting the information that goes into the model and understand how the information they helped gather influences the outcomes of the models.

Before, it was government coming out and saying, "This is what we are going to do, based on X." Well, X was different from what fishermen thought they knew and different from what they saw on the water. Now, we participate in the deliberations of science. For example, the Atlantic States Marine Fisheries Commission manages shrimp. Fishermen are on the advisory boards, involved in the shrimp research; they work with the technical committees looking at the different year-classes. And this past year, we went further than we ever have before. The industry advisory board and the scientific committee worked together prior to making recommendations to the full commission. The two committees came up with a common, unified recommendation.

Before, we were setting ourselves up for failure, for a fight, for a lawsuit. Now, we work together and came up with something.

In closing, what are the social, cultural, and economic impacts of cooperative research and management?

I believe they include

- building a common language we can all speak;
- fostering our understanding and appreciation for each other—it is a place to listen to others, to learn about others;
- providing the setting for professional growth and evolution in fishermen, scientists, and managers—this is particularly exciting for the next generation of fishermen, scientists, and mangers;
- building broader cooperation throughout the industry;
- providing the information needed for tailored effective management measures for particular sectors, particular communities, and particular ecosystem dynamics;
- supplying economies of scale—getting a bigger bang out of our research and management buck;
- building credibility and legitimacy in the science and the management—we can build trust;
- supplying the potential to engage conservation organizations more constructively; and
- letting us be visionaries, thinking long term.

I look forward to participating in this next panel's discussion and in the entire symposium because I believe cooperative research and management have the potential to fundamentally transform the way we all do what we do.

American Fisheries Society Symposium 52:87–93, 2006

Session 1B: Making a Difference by Working Cooperatively

Social, Cultural, and Economic Impacts, Panel Summary

Troy W. Hartley
Northeast Consortium
Department of Resource Economics and Development
University of New Hampshire
210 James Hall, Durham, New Hampshire 03824, USA

> Cooperative research can lead to the development of common terms of reference, broadening both parties thinking about resource functions. New understandings emerge when communication is good, interaction is direct, and both parties recognize…a shared definition of an existing problem. [Linda Behnken, executive director, Alaska Longline Fishermen's Association]

The first session of the joint Sea Grant/American Fisheries Society symposium focused on the impacts of cooperative fisheries research and management. Troy Hartley with the University of New Hampshire and the Northeast Consortium moderated the panel on social, cultural, and economic impacts. Below are some speaker highlights, followed by a summary of the facilitated panel-audience discussion. White papers prepared by the keynote and panelists follow this panel summary.

Keynote speaker Pat White, lobsterman, CEO of the Maine Lobstermen's Association, and former Pew Oceans Commissioner, launched the panel's discussion with what he has seen as the significant impacts of cooperative research and comanagement:

- Builds common language among scientists, fishermen, and managers;
- Fosters an appreciation and understanding of each other by providing a safe place to listen and learn from each other;
- Enhances industry capacity, with long-term implications through the professional growth of younger fishermen;
- Promotes broader cooperation and understanding between industry sectors;
- Provides the necessary information for locally tailored, effective management measures for specific sectors, communities, and ecosystem dynamics;
- Demonstrates economies of scale that leverage research and management capacity;
- Builds credible, legitimate science and management and, in turn, greater trust; and
- Provides opportunities to engage conservation organizations, and promotes more visionary thinking that considers the long-term health of sectors, communities, and ecosystems.

Pat White, lobsterman, CEO of the Maine Lobstermen's Association

The panel, rich in fishing, coastal community, and research experience, presented and discussed several of these themes:

- Cooperative research has made business sense (financial impacts on the bottom line, knowledge to become more efficient at catching fish).
- Sparks his inquisitive nature; he asks more questions now than he did before becoming a cooperative researcher. Why do we fish this way? What is really going on with his gear? He is constantly discussing new ideas with research partners and scientists.
- "I had never engaged in this type of thought prior to taking part in cooperative research."

Vincent Balzano, FV *North Star*, Portland, Maine

Lynn Palensky, sub-basin planning coordinator for the Northwest Power and Conservation Council

- Considerable flexibility and communication within a consistent, clear, and organized management regime (e.g., schedules, budgets) could result in effective coordination across 58 varied subbasins.
- A comanagement program on this scale faces considerable challenges (e.g., funding, infrastructure, procedural and technical guidance, schedule, and communication).
- Providing regional stakeholders incentives to participate and provide the plan, giving people a feeling of inclusion and pride and providing technical guidance and assistance helped overcome these challenges.

Linda Behnken, director, Alaska Longline Fishermen's Association

- All boat sizes and the full demographic of a fleet should engage in research to avoid skewing results and subsequent management measures toward the larger boats that can carry larger scientific research teams.
- Breadth of involvement among communities provides data that are otherwise unavailable and facilitates more precise management measures with broader effectiveness throughout the fleet.
- Communication and coordination processes of cooperative research and comanagement are complicated and often are improved with facilitators or coordinators.
- Can lead to more enforceable, implementable rules and greater trust among partners.
- Slow or delayed action by regional and federal partners, particularly National Marine Fisheries Service, to address issues that are beyond the control of local community partners has undermined the newly built trust.

Based upon these opening remarks and insights, the panelists, keynote speaker, moderator, and audience members raised several critical issues, particularly related to participation and the nature of representation in cooperative research and comanagement. Based upon the panelists' preparatory white papers, they also raised critical questions about trust-building and what working cooperatively may achieve—these remain important issues and questions for future research and discussions.

Participation

There was considerable discussion and no consensus among the panelists and the audience about participation in cooperative research and comanagement—who participates, are some more entitled than others, whose responsibility is it to ensure that all stakeholders are aware that decisions are occurring?

Several panelists noted that cooperative research and comanagement need committed individuals willing to invest their time and energy—"keep a diversity of perspectives, but you really want those people with a lot of energy and that are very engaged." Others countered that involving only "high liners" or those predisposed to participating poses the risk of skewed data and findings that are not representative of actual industry activities and impacts. A panelist suggested "get all types of fishermen from the fleet involved, then we bring more people to the management forum, educate them, involve them, and most importantly, you ensure you are not biasing the data." Committed participation extends to an acceptance of the cooperative process and its outcome. A panelist noted how trust and credibility of the process itself are undermined when participants elect to avoid the cooperative process and go

Andrew Day, executive director, West Coast Vancouver Island Aquatic Management Board

- Representation, funding, and leadership styles can go through periods of change and stability.
- Unique emergent characteristics of a process as the cooperative management experience unfolds and matures within the existing social network among people.
- Cooperative management experiences persistent organizational challenges (e.g., funding, how to prioritize, and administrative activities).
- Cooperative management occurs within the existing political context, and thus, the change that cooperative management presents will face resistance and produce conflict.
- Building and maintaining cooperative management takes highly committed people, willing to get involved. Attending to the underlying motivation for individuals' involvement is important to help participants work through the political and organizational challenges and the changes.
- Cooperative management is a significant shift in the way things are done, emphasizing codependence and a different way of thinking and interacting with one another.
- Sustaining cooperative management in the long term has proven extremely challenging in other contexts.
- Inability to sustain cooperative management programs has contributed to making many participants cynical about them.

directly to NMFS, an individual fishery management council member, or Congress to address their interests; "nothing compromises the trust in the system, to everyone involved, more than this."

It was noted that cooperative research and comanagement often includes only those who usually participate, with a panelist arguing that it would increase the overall understanding of the industry and the impacts of regulations if these less frequent participants engaged in the process; "We are missing the ones that do not usually participate in the management process... If we reach out to those people, it would strengthen the understanding within industry." However, a fisherman noted that many of these individuals are not passive or awaiting an invitation and that he has been "equally frustrated with those that have resisted the process and stay out of it." A native communities representative added, "a lot of times responsibilities and roles to manage resources [in small native communities] are handed down within families. So, you need to be aware of how things work in these communities when looking at choosing representatives." Furthermore, a New England participant noted that conservation organizations and the recreational fishing industry are not actively participating in cooperative research and comanagement in New England.

It was not easy to define the appropriate, adequate representation or balance between committed participants willing to do the work and invest the time to make cooperative research and comanagement work with those new voices that have elected not to participate, but could enrich the overall understanding of the industry and the fishery. A panelist noted that cooperative re-

- Defines fishermen's knowledge as anecdotal and localized, since it is the qualitative observations and views gained through the experience of fishing.
- Scientific knowledge is based upon research methods and has been more readily accepted in the science and policy process.
- Defines fishermen's knowledge as anecdotal and localized, since it is the qualitative observations and views gained through the experience of fishing.
- Scientific knowledge is based upon research methods and has been more readily accepted in the science and policy process.
- Cooperative research presents the opportunity to integrate fisheries-based and research-based knowledge and produce a new kind of data or new knowledge.
- Cooperative research may be a means to bring these two types of knowledge together, although to realize its full integrative potential more varied participants and fishermen and additional opportunities beyond chartered vessels might be needed.
- Incorporation of cooperative research results into management has lagged behind the number of cooperative research projects undertaken.
- Fishery management councils are working on procedures to incorporate results into the management process, peer review remains critical.

Teresa Johnson, Ph.D. student, Rutgers University

Bonnie McCay, anthropologist, Rutgers University

search and comanagement are very democratic processes, although "sometimes fishermen are independent to their demise." The panel's and audience's struggle with the representation issue reflected the tussle between desirable ideals and pragmatic needs. Other panelists concurred—it is important to "aim high" and provide opportunities to participate, along with doing things to get buy-in to the cooperative process:

- Provide purpose and incentives for participation, including resources, a clear opportunity to be heard and make a difference, and salience to participants.
- Involve as many perspectives as possible in defining the problems and issues at hand; this will help ensure that those interested stakeholders will participate.
- Clearly present meeting goals, publish purpose, agenda, and location of meetings well in advance, and ensure that meetings are comfortable and not intimidating.
- Invest time on building understanding and reaching consensus.

- Safeguard the integrity of the outlined cooperative process, so that the process is fair and participatory.

Improving Understanding and Trust

Panelists discussed the use of data and its role in generating understanding and trust. Industry representatives expressed a concern that information and data they collect and share are disregarded or not used by scientists, misused, or used in unexpected ways. The panelists also addressed how to proceed when there are concerns that information and data may be perceived as potentially damaging to a stakeholder's interest. Steps panelists have taken to address these concerns involve openness, honesty, and clarity and have included

- prioritizing existing information gaps and research needs, while acknowledging that you cannot control what the results and findings will be or where the data will go once it is out and public;
- describing at the start and in a detailed formal manner (e.g., document) how information will be (and will not be) used and how it might be protected;
- using trusted independent consultants and third parties to manage data, if necessary; and
- agreeing that data will not be used unless all stakeholders agree upon its use.

Panelists found that these types of activities generated clear, formal, and upfront expectations. Nonetheless, there will be times when research findings do not support a particular stakeholder's position or interests. Communication plays a central role in achieving this clear, shared understanding of expectations and how the outcomes will be used, independent of whether the outcomes are good or bad for a particular stakeholder's position.

When clear expectations are not shared or there are changes in how people thought information was to be used, it can breed mistrust. A panelist noted, "Communication is key…keep communication open and keep people informed, otherwise rumors will build." In his experience, the panelist tried not to rely on trust, but rather to "supplement trust with a formal (written) agreement." The formal agreement also helps ensure consistency when there is turnover among participants. One fisherman noted, "Distrust in New England is very real. They (fishermen) were required to submit trip reports for landings from all fishing trips. They were told that this information was not going to be used to hurt them and that it would help them. … then there were rolling closures implemented, mandatory closures…. They saw that all the places where they caught fish were the areas that were closed… [managers] need to be honest… Distrust is there for a reason."

Knowledge Integration

One panelist claimed that better knowledge integration and, particularly, stronger linkage between cooperative research and management would result in less "frightening and draconian" management decisions. A strengthened linkage between cooperative research and management would integrate knowledge. A fisherman noted that understanding the management process, and taking the time to understand those priorities, makes for a more effective cooperative partner. One fisherman in the audience added that it is important to know, "how it [cooperative research findings] is being put into the management system, how did you get it integrated."

Future Social, Cultural, and Economic Questions—Next Steps and Concluding Remarks

The panel discussed many of the critical social, cultural, and economic issues at play in cooperative research and management—participation, information and data use, trust, incentives, and understanding, communication. Further hypotheses emerge from the collective writings, presentations, and discussion among these panelists and the audience. The concept of trust stands out—how is it built, how is it destroyed, and what difference does it make? A panelist noted that "trust is fragile," a theme that repeated itself in several of the panelists' white papers. Common themes emerged describing a sequential process toward building trust that was dependent upon significant time, communication, and direct interactions. Fishermen, scientists, and managers do not speak the same language. But at sea, in planning meetings, or other nonmanagerial settings, there is more casual, open conversation and more opportunities to informally and less contentiously exchange information and communicate.

Panelists described how trust was built over time, through stages of development. As prerequisites, the issue at hand has salience often with management implications that establish incentives to form partnerships, the expectations of a partnership are clear, and research parameters defined. Nonetheless, participants remained optimistic for the future. Cooperation in research and management can produce social, cultural, and economic benefits for fishermen, fishing communities, governance, and society.

American Fisheries Society Symposium 52:95–97, 2006

Making a Difference by Working Cooperatively: One Fisherman's Perspective

Vincent Balzano
FV North Star
31 Vines Road, Saco, Maine 04072, USA

Why Do Fishermen and Scientists Participate in Cooperative Research or Management?

It is important to note from the beginning that this opinion paper represents the ideas of one fisherman. I do not speak for fishermen as a whole, as each of us makes decisions for our own reasons. Fishermen are independent, resilient, and hardworking. To try to group them all into one category together is difficult and does not always provide an accurate representation. There is no textbook for the industry, but rather what works for each individual fisherman and his business. Most fishermen inherit their fishing skills or learn them on their own through trial and error. Fishing is my culture; I was born into a fishing family. As a child, I thought this was the way of life for everyone. However, I *can* tell you why *I* got involved in collaborative research and what motivates me to stay involved.

The first project I ever worked on was the original grid work for the whiting fishery in the Gulf of Maine from 1995 to 1999. There was no direct monetary compensation, but it was permitted as an experimental fishery so you could keep the catch. We were also required to take an observer on board. The project was started because all small mesh fishing was shut down in the Gulf of Maine and we needed to prove to the government that we could catch whiting without catching groundfish. At the time, I really was not aware that I was involved in a "cooperative research project" as I was just following the requirements to prosecute the fishery. I worked with Dan Schick from the Department of Marine Resources, and it was a very positive learning experience for me. I was introduced to the experimental fisheries permitting process, sea sampling techniques, and data gathering protocols. More importantly, it became the stepping stone that would expose me to the fisheries management council process.

"...there are economic incentives to doing cooperative research. The opportunity to work on cooperative research has added some stability to my income stream during a difficult time for groundfish fishermen in New England."

There are many reasons why I have remained interested in cooperative research since my first exposure in the mid-1990s. For one, there are economic incentives to doing cooperative research. The opportunity to work on cooperative research has added some stability to my income stream during a difficult time for groundfish fishermen in New England. I am also excited about adding to my knowledge of fishing and what is really going on with the gear and how it works. As a third-generation fisherman, fishing is instinctive. Research was the first time I thought about *why* I did what I do. By participating in collaborative research, I have an opportunity to understand fishing better and gain a different perspective. To actually learn something—you can't put a price on knowledge.

Many of the projects I have been involved in are testing new fishing gear or fish gear in new ways. What I am really doing is getting paid to play and test gear—what could be better? In the long run, doing research helps me become better and more efficient at catching fish. I see it as a win for me and a win for the researchers. I get to learn new gear designs and see how my gear actually works, and the researcher gains insight into how fishermen will, in reality, use a piece of new gear on the water. We both gain something by working together.

Many Have Stated that Cooperative Research and Management Improves Understanding and Trust between Fishermen and Scientists, but How?

In my case, collaborative research did not actually change my relationship with scientists that much. I did not have many thoughts about scientists before I began doing research. When you start the development of a project from the proposal stage to the fieldwork to the development of the data, you spend a lot of time with the scientists. You realize that they are committed to what they are doing. It is hard to discredit them and not consider their opinions. If you have the opportunity to work beside someone and have a relationship that is not confrontational, then it helps when there is something difficult to discuss. Collaborative research allows me to work with scientists on a different level with other management issues aside.

Are Fishermen's and Scientists' Knowledge Integrated, and Do New Understandings Emerge? If So, How?

As I mentioned, most of the projects I have worked on are gear projects. If any of the results are actually going to be used, the fishermen need to buy into and support the use of the gear. If it works, they will use it even if they are not required to under any regulation. The real driving force is economic success from using the gear. But, you can't catch what is not there, so it is important to know what is really going on out there. Is the gear not catching the fish, or are the fish not there? That is why gear experiments usually have a comparison tow, so you know what is out there while you are testing the new gear.

In order to be an effective partner in a project, you need to have a foundation of fishing. I think that is an important component that often gets overlooked. It is not enough to have a "fishing industry partner" and check the box for those criteria in a proposal. You need to involve a fisherman who has knowledge and experience that can add value and insight to the project because they have fished under many different conditions, have conducted their own experiments over the years, and understand how their gear works.

Getting both partners involved at the beginning of the proposal process develops true collaboration. While I am out conducting the research for a project, I always have a million more questions about what is happening and why. I am constantly discussing new ideas with the project partners and trying to think of new ways of doing things. Each new project seems to stem from ideas I discuss with the scientists. The projects are a continuation of what we have learned together. We are trying to find a better way of harvesting.

"If any of the results are actually going to be used, the fishermen need to buy into and support the use of the gear. If it works, they will use it even if they are not required to under any regulation. The real driving force is economic success from using the gear."

One of the best experiences I have had with cooperative research is the opportunity to visit the flume tank at Memorial University in Newfoundland. The flume tank simulates conditions at sea while a net is in the water. You can observe how the gear is working—first hand. I traveled to Newfoundland with a group of fishermen and scientists to test different types of gear. It was great to watch the gear underwater and discuss and exchange ideas with the scientists. Together, we learned a lot and discussed a lot of new ways to test the gear. By traveling and testing the gear together, we were all learning from the other. So, to answer the question: yes. My experience has been that fishermen's and scientists' knowledge are integrated when doing collaborative research. New ideas emerge when more than one person is thinking about how to do something. This is the case with cooperative research projects.

What Critical Social, Cultural, and Economic Impact Questions Should We Be Thinking about as the Symposium Considers Lessons Learned and Future Opportunities for Cooperative Research and Management?

Cooperative research takes time. I think it is misleading to believe that research is easy money. You have to be involved from the beginning to make a project successful. It takes a lot of meetings and preparation prior to even getting on the water and additional time to discuss the results. Going to the meetings and participating is part of doing business. Also, fishermen are generating many of the research questions. Many times, the research is focused on testing ideas brought up by the fishermen.

Finally, as more restrictions are implemented on the fishing industry, cooperative research is the only way of regaining lost access. I recognize that participating in research is not for everyone. That's okay. My hope is that during these challenging times we can keep the dialog open and constructive so that we can all benefit from the conservative measures that have been taken.

American Fisheries Society Symposium 52:99–101, 2006

Perspectives of a Cooperative Fisherman

LINDA BEHNKEN
Alaska Longline Fisherman's Association
123 Riggs Road, Sitka, Alaska 99835, USA

Introductory note: In addressing the questions below, I draw on my experience, limited research, and discussions with both fishermen and scientists. I consider my list of answers to be accurate and representative, but certainly not complete or all-inclusive.

Why Do Fishermen and Scientists Choose to Participate or Not Participate in Cooperative Research or Management?

Fishermen choose to participate or not participate in cooperative research and/or management for a number of reasons. Those who choose to do so generally cite one or more of the following reasons:

1. A recognized data need that consistently leads to mismanagement or underestimate of stock abundance;
2. Awareness of a looming problem that could effect their fishery;
3. Financial compensation commensurate with the time and effort demands of the project; and
4. Awareness that the scale of a recognized problem is ill-suited to the existing regulatory regime and could be better resolved at the local level.

Fishermen choose not to participate in cooperative research and management if they

1. generally mistrust the management system or mistrust scientists or managers engaged in a particular project;
2. receive little or no benefit or compensation for the time and/or space needed to participate; and
3. believe the data they generate will not be used.

Scientists choose to participate if they

1. recognize a means of collecting otherwise unavailable or unaffordable data or improving their understanding of ecosystem relationships and
2. recognize that including fishermen in research and management will improve communication, cooperation, and support for management decisions.

Scientists choose not to participate in cooperative management when they

1. consider data generated by fishermen to lack objectivity or rigor;
2. are unable to build relationships with fishermen that could lead to cooperative work;
3. lack funding to engage fishermen in cooperative work; and
4. fear expectations on the part of the involved fishermen regarding outcomes could result in hostility and mistrust.

Many Have Stated that Cooperative Research and Management Improve Understanding and Trust between Fishermen and Scientists, but How?

Cooperative research and management can improve understanding and trust between fishermen and scientists when expectations, research parameters, and management implications are carefully discussed and specified in advance. In other words, scientists and fishermen that enter into cooperative research projects with a poor understanding of each other's perspective may emerge with more mistrust and misunderstanding simply because expectations were not met. However, good communication prior to and throughout a cooperative project can dramatically improve the understanding on the part of fishermen of the objectivity, systematic approach, and scientific rigor practiced by good scientists and impress on scientists the depth of ecological knowledge held by many fishermen as well as the practical realities of working at sea. Both often come to recognize the overlap of shared concerns and objectives and to appreciate the constraints under which each other work.

Also, critical to building trust and understanding is direct interaction. Working together, particularly at sea or in a less structured setting (i.e., outside an office) allows for open discussion and an exchange of ideas that is unlikely to happen during phone conversations, testimony, or other indirect means.

Are Fishermen's and Scientists' Knowledge Integrated, and Do New Understandings Emerge? If So, How?

Integrating the knowledge of fishermen and scientists is often difficult given the differences in language—technical and quantifiable versus descriptive, anecdotal, or historic. Cooperative research can lead to the development of common terms of reference, broadening both parties thinking about resource functions. Again, new understandings emerge when communication is good, interaction is direct, and both parties recognize shared goals or a shared definition of an existing problem.

What Critical Social, Cultural, and Economic Impact Questions Should We Be Thinking about as the Symposium Considers Lessons Learned and Future Opportunities for Cooperative Research and Management?

Cooperative management provides avenues for improved fisheries management by enhancing data collection and resource understanding, strengthening the support of fishermen for management decisions and regimes, and allowing for scale and locally appropriate policy. Realizing these benefits depends on clear communication about project objectives and the expectations of involved parties before and throughout a project. Directly addressing the issue of trust, how it is developed and how it is broken, will encourage all parties to air concerns. Scientists need to be aware that for most fishermen that trust is fragile; it is slow to build and quick to disappear. From the fishermen's perspective, the scientists are undertaking well-paid experiments while the fishermen are placing their livelihoods at stake. Cooperative management and research offers tremendous opportunities but is not without risks.

That said, cooperative research should serve as an addition to, not a substitute for, traditional research. Some segments of industry are better equipped to participate in research than others, which may bias data collection and, in turn, fishery managers. Scientists and managers should continue to develop relationships with fishermen that allow scientists access to commercial vessels as research platforms as well as allowing fishermen to provide data collected without scientists aboard. This will help prevent a bias toward large vessels as partners in cooperative projects.

“Cooperative management provides avenues for improved fisheries management by enhancing data collection and resource understanding, strengthening the support of fishermen for management decisions and regimes, and allowing for scale and locally appropriate policy.”

One of the avenues of expanding cooperative management that seems of particular importance is community-based fishery management. Community issues related to resource access are frequently on a scale that defies traditional management boundaries. Likewise, management regimes often favor aggressive and well-financed fishing operations or corporations at the expense of small-scale, community-based operations. Communities empowered to develop an inclusive process informed by scale-appropriate data can develop resource management solutions that would elude outsiders or result in a regulatory regime far too complex to be enforced without community support. Likewise, community-based management is likely to lead to management regimes appropriate to the size and gear-type utilized by local vessels, hence maintaining access opportunities for community residents. Since many coastal residents have few alternative employment or life-style opportunities, protecting this access to local marine resources is critical.

There is growing recognition both in Alaska, Canada, and the lower 48 of the need to encourage community-based management as a means of protecting the viability of coastal economies. Given the growing complexity of fisheries management and the need to incorporate into management decisions, the combined wisdom of multicultural entities, community-based management should be supported by making data available as needed and providing communities with logistical support.

Key Questions for Future Thought

- How can the information and wisdom generated by fishermen be incorporated into ecosystem management?
- How can scientists ensure that data generated by fishermen is used and not dismissed as anecdotal or unquantifiable?
- How can fishermen and scientists develop a common language and approach that satisfies both parties and generates useable data?

American Fisheries Society Symposium 52:103–105, 2006

Impacts: Making a Difference from Working Cooperatively—Social, Cultural, and Economic Impacts

Andrew Day
WCVI Aquatic Management Board
3017 3rd Avenue, Port Alberni, British Columbia V9Y 2A4, Canada

Miscellaneous Thoughts Regarding the Discussion Questions

- It is more enlightening to ask the individuals involved in a project than to assume that you understand what is motivating them. In fact, understanding what motivates each partner, establishing a common intention, and revisiting these in the face of challenges are key elements of successful collaboration.
- Collaboration is heavily influenced by its political, cultural, economic, and ecological context.
- Good fishermen are good scientists. Good scientists are good fishermen.

Why Do Fishermen and Scientists Choose to Participate in Cooperative Research or Management?

Money

Fishermen, or their associations, are increasingly expected to share more management costs. Many are looking for ways of doing this cost effectively by partnering with academic scientists, developing more cooperative working relationships with government scientists, or hiring a scientist/consultant to work for their association. In all these cases, some fishermen, or their association representatives, are looking to have more say in the science.

Scientists may have more success getting funding if they can show partnerships. Some believe that cooperative relationships have a better chance of producing innovative methods and outcomes that will result in marketable products or models.

> “...understanding what motivates each partner, establishing a common intention, and revisiting these in the face of challenges are key elements of successful collaboration.”

If a scientist is spending fishermen's funding on science, and reporting to them, they may be more cooperative in the application of the funds.

Effectiveness

Fishermen, or their association representatives, might be interested in new, innovative approaches that reflect their theories or knowledge. If they do not believe the past or current approach is effective, they may look for others who will incorporate their knowledge or ideas.

Some scientists believe that the more cooperative and close the working relationship with fishermen, the greater likelihood that their research results will be more innovative, practical, understood, and accepted.

Values

Some fishermen and scientists believe in cooperation and value it in and of itself or are motivated to cooperate because they enjoy working in groups, learning from others, testing theories in a practical setting, and so forth.

Power/control or validation

Some people like to work cooperatively or in groups because they want to be the leader, desire social status, or want to play a role that makes them feel validated or because they feel they can have more control over the outcome.

Why Do fishermen and Scientists Choose Not to Participate in Cooperative Research or Management?

- They are getting what they want and need from the status quo.
- There is no incentive to do so.
- They do not like working with other people.
- They have had some bad experiences.
- They believe it will take more time or cost more money than it will bring benefits.
- They are not organized enough.
- It is not a high priority on the to-do list.

Many Have Stated that Cooperative Research and Management Improve Understanding and Trust between Fishermen and Scientists, but How?

I have seen cases where scientists or managers had never actually seen the place where harvesting occurred or how it occurred. When they did see it, it totally changed their understanding of the fishery, the stocks, and fishermen. Similarly, I have seen fishermen start to understand how difficult and valuable it can be to test theories rigorously and provide sound advice to decision makers by participating more directly in science and decision making.

I believe this understanding builds through experience through direct interaction. It is a combination of body language, verbal communication, and context. Part of it is recognizing the other person as a human being, part of it is having the opportunity to clarify misconceptions and dispel past beliefs or biases, and part of it is the tendency to harmonize differences in a group. It takes a lot of time in some cases, especially where past mistrust or prejudices are high or where direct interactions are infrequent. Some people never really get past their prejudices.

> “I have seen fishermen start to understand how difficult and valuable it can be to test theories rigorously and provide sound advice to decision makers by participating more directly in science and decision making.”

Trust is fragile. It takes a long time to build and is based on people's actions much more than their words. It can be broken very quickly. I have worked with groups where we have produced an evolving chart of actions that build trust and actions that undermine them. This seems to help bring the issue of trust more out into the open and provides a way to talk about actions undermining trust in a constructive way.

Are Fishermen's and Scientists' Knowledge Integrated, and Do New Understandings Emerge? If So, How?

Sometimes knowledge integrates, depending largely on the people involved and the extent to which they work well together and are open to each other. It is an organic process of exchange and fusion. The chemistry between individuals is key.

I have seen other circumstances where a fishermen is simply expected to be a technician for the scientist or where the fishermen has no time for the “egghead's” knowledge, and so there is little exchange or integration.

What Critical Social, Cultural, and Economic Impact Questions Should We Be Thinking about as the Symposium Considers Lessons Learned and Future Opportunities for Cooperative Research and Management?

1. How can partnerships regarding one species or use be integrated in an ecosystem approach?
2. How are the values of fishermen and scientists changing? What is driving the change? What are the possible outcomes of any detectable trends?
3. Are the ownership and benefits of access to fisheries being increasingly consolidated? If so, is this resulting in greater control over decision making and research agendas, or is it resulting from them, or both, or neither?
4. "The Nuu-chah-nulth phrase 'Hishukish Ts'awalk' (everything is one) embodies sacredness and respect. Nothing is isolated from other aspects of life surrounding it and within it. This concept is the basis for the respect of nature that our people live with, and also contributes to the value system that promoted the need to be thrifty, not to be wasteful, and to be totally conscious of your actual needs.... With respect for other life forms comes the Nuu-chah-nulth belief in the spirituality and sacredness of life and the earth, and in a oneness between humans and the environment." [Roy Haiyupis, Ahousaht First Nation]

"Are the ownership and benefits of access to fisheries being increasingly consolidated? If so, is this resulting in greater control over decision making and research agendas, or is it resulting from them, or both, or neither?"

Where I work, First Nations teachings state that research and management are part of an on-going relationship between people and other beings. The core of the relationship is a value-system rooted in a belief that all life is interconnected and sacred. This belief and value system leads to actions based on respect. These actions include different approaches to research and management than those practiced under the current dominant paradigm.

Often fishermen or First Nations' knowledge is expected to fit within the western scientific paradigm and the values it is based on. This can result in diverse approaches, knowledge, and wisdom being lost.

Is it possible to reconcile the research, use, and management methods that result from this value and belief system with those practiced under the current dominant paradigm? If so, how?

American Fisheries Society Symposium 52:107–110, 2006

Subbasin Plans: Guiding Fish and Wildlife Recovery

Lynn Palensky
Subbasin Planning Coordinator, Northwest Power and Conservation Council
851 SW 6th Avenue, Suite 1100, Portland, Oregon 97204, USA

Introduction

The Northwest Power and Conservation Council (Council) recently completed one of the largest locally led watershed planning efforts of its kind in the United States, an effort that resulted in separate plans for 58 tributary watersheds or mainstem segments of the Columbia River. The plans identify priority restoration and protection strategies for habitat and fish and wildlife populations in the United States portion of the Columbia River system. Planning occurred in far-reaching parts of Oregon, Idaho, Washington, and Montana of the vast Columbia River basin. Various local planning groups, recovery boards, and Indian tribes developed the plans, including Canadian entities where the plans address transboundary rivers. The effort spanned 2 years of intense administrative process, including independent scientific review and public scrutiny of the plans. The plans were completed on time and under budget. The next steps are to implement the plans over the next several years based on priority areas and populations identified in the plans.

Background

The Columbia River Basin Fish and Wildlife Program directs more than $140 million per year of Bonneville Power Administration electricity revenues to protect, mitigate, and enhance fish and wildlife affected by hydropower dams. Many types of projects implement the fish and wildlife program. These include improving and protecting habitat, improving fish passage at dams, rebuilding naturally spawning fish populations through the careful use of hatcheries, and researching factors that affect fish and wildlife survival. This enormous amount of work is now coordinated through strategic plans for each major tributary and distinct main-stem segment of the Columbia River. In 2004 and 2005, the Council adopted the 58 subbasin plans into the fish and wildlife program to provide direction for projects and project-funding decisions in the future. The plans were developed collaboratively by state and federal fish and wildlife agencies, Indian tribes and watershed councils with Bonneville funding, and Council guidance. Subbasin plans ensure that projects to implement the program are planned and evaluated principally at the local level. Subbasin plans direct fish and wildlife and water quality management activities funded by Bonneville and also provide the context in which the Independent Scientific Review Panel reviews project proposals for funding through the program. Subbasin plans integrate strategies and actions funded by others, thus ensuring that each plan serves the Council's purposes under the Northwest Power Act and also accounts for Endangered Species Act and Clean Water Act requirements, and other laws governing natural resource management, as fully as possible.

“The plans identify priority restoration and protection strategies for habitat and fish and wildlife populations in the United States portion of the Columbia River system. Planning occurred in far-reaching parts of Oregon, Idaho, Washington, and Montana where the Columbia River and all its tributaries made its home.”

Each subbasin plan includes the following components:

- *A subbasin assessment*, which provides a description of historical and existing conditions, past efforts to improve fish and wildlife survival, an assessment of limiting factors, and the biological potential of the subbasin. The assessment also identifies protection and restoration opportunities.
- *A strategic plan*, with a 10- to 15-year horizon, describing the vision and biological objectives for the subbasin and the strategies to be taken to achieve the biological objectives. The strategic plan includes a comprehensive monitoring and evaluation plan.
- *An implementation plan*, with a 3-year time horizon, detailing the specific actions and measures to implement the strategies in the subbasin plan. Projects are funded for 3 years at a time.

How Did We Get Here?

Need for better science and review

Prior to adoption of the 2000 Fish and Wildlife Program, the Council conducted an annual solicitation for projects to implement the program. The work completed in the basin represented a collection of projects or measures designed to restore fish and wildlife, but they were not well connected in a scientific or priority sense. An independent scientific review of the previous version of the fish and wildlife program criticized it for not having a strong scientific foundation. The scientists said such a foundation is essential to build a strong program that knits together priorities and needs on a scale longer than the local watershed. There were other problems with this approach, as well. Soliciting 400–500 projects per year limited the amount of time and effort that went into the scientific review by the shear volume of proposals to review. The project sponsors also endured hardship in gearing up for a major proposal each year when they really needed to engage in multiyear funding to focus time on implementation.

In response to these criticisms, the Council began a comprehensive revision of its fish and wildlife program. First, the Council amended the program by adopting a three-tiered framework of vision, objectives, and strategies at three different geographic scales (basinwide, ecological province, and subbasin), tied together with a consistent scientific foundation. The Council also adopted basinwide provisions and described how it proposed to add more specific objectives and measures to the program through integrated subbasin plans for the tributary subbasins of the Columbia and for specific mainstem reaches. Subbasin plans will guide the expenditure of Bonneville funding to the most appropriate places and in the most effective ways. The recently adopted subbasin plans describe the vision, objectives, and strategies at the smallest geographic scale described in the program.

“The project sponsors also endured hardship in gearing up for a major proposal each year when they really needed to engage in multiyear funding to focus time on implementation.”

The next steps in the program are to define the provincial goals, objectives, and strategies and, finally, to implement strategies at the subbasin and provincial scales where they are appropriate. The 2000 program also describes a rolling provincial review of projects on a 3-year cycle. This allows a focused effort on prioritization and funding in a smaller geographic area over a 3-year cycle. The most important feature of this process is the ability for the independent scientists to increase review time and depth for individual proposals, resulting in a more rigorous review. Final project recommendations then come with more certainty and credibility.

What the Plans Represent

On May 28, 2004, the Council received recommendations for program amendments in the form of subbasin plans. The Council received 58 plans from various planning entities and made those recommendations available for public review and comment, including review by a team of independent scientists. The plans represent scientifically based and reviewed locally developed documents that can guide project funding to priority areas and activities. In general, the plans will and can

- coordinate and focus projects,
- guide Bonneville Power Administration investment,
- guide federal plans and other investments,
- serve as a basis for ESA recovery plans,
- provide ESA conditional compliance, and
- provide a basis for monitoring and evaluation.

Scientifically supported plans

The foundation for all the subbasin plans began with the development of a document entitled *A Template for Subbasin Assessment*, which identified a range of analytical tools and data types that would be useful for subbasin planning. Using the template as a foundation, the Council developed the *Technical Guide for Subbasin Planners* to help ensure that plans had a consistent format and content. Assessment data that had been collected across the basin at a consistent scale was provided to the planners to use in developing their technical assessments. This is the guidance document that planners used to develop each piece of the plan. The assessment in the plans was to be based on existing information. The data and information were plugged into one or more modeling tools provided by the Council and NOAA Fisheries to determine the limiting factors and the areas with greatest opportunity for restoration and protection. Out-of-subbasin assumptions such as juvenile fish survival, main-stem harvest, and ocean survival were considered. Technical assistance was provided to planners at the province and regional levels as well as from NOAA's technical recovery teams.

> “The Council took a flexible approach in the process in order to accommodate the various existing entities and ongoing watershed and recovery planning activites in each state and subbasin.”

Locally developed and supported plans

In August 2002, the Council solicited recommendations for amendments to the program (subbasin plans) at the subbasin level from the region's state and federal fish and wildlife agencies, Indian tribes, and others, as required by the Northwest Power Act. The Council also worked with the Bonneville Power Administration to secure funding support for planning groups, the first time that funding has been made available to help develop fish and wildlife program amendment recommendations. The Council took a flexible approach in the process in order to accommodate the various existing entities and ongoing watershed and recovery planning activities in each state and subbasin. The states of Idaho, Montana, Oregon, and Washington each had a statewide subbasin planning coordinator funded through the program to help define and guide the process within their state and facilitate communication with the Council.

Each planning group, existing or new, developed a process by which it would develop and submit the plans. Contracts were let to entities within each subbasin for a specific scope of work—sometimes multiple contracts within a subbasin and, other times, one contract for several subbasins. The Council signed more than 90 contracts with tribes, the state fish and wildlife agencies, conservation districts, consultants, and fish recovery boards to complete this work over 2 years. Most groups formed planning and technical teams, each with different functions. The assessment work started with the technical team, and the planning team had the review and writing functions. Many groups also created public outreach teams as well to engage the local community and hold public meetings and hearings.

In general, the plans had buy-in from tribes, counties, the state agencies, legislators, and others with interests in the subbasin.

The Council, as a matter of Northwest Power Act requirements, held two separate public review and comment periods in the process as well. The first began immediately after the plans were submitted, and the second was just prior to adoption by the Council. The plans were also vetted through statewide coordination groups and at a regional level through the Regional Coordination Group, created specifically to facilitate discussions of subbasin planning at a regional scale.

Next Steps and Opportunities

The Council adopted the management plan portions of all 58 subbasin plans into the fish and wildlife program by June 2005. The underlying technical assessments and inventories will be placed in an appendix to the program. The plans already are helping agencies deal with issues regarding fish and wildlife, water, land, hydropower, and the Endangered Species Act. The plans helped to build better relations and deeper understanding with the public about complex and controversial environmental issues. The plans will continue to promote improved, collaborative planning consistent with local and regional goals and the direct limited funding to fish and wildlife projects with the highest probability of success. Our work is really just beginning. The next step is to implement the priorities identified in the plans in a timely fashion.

There are other steps that the Council is taking that are necessary to make this process successful and meet our program objectives. The program calls for provincial objectives as well as at the larger basinwide and smaller subbasin level. In order to do this, the Council must take on the task of integrating or "rolling up" subbasin plans to the provincial scale. Another modeling tool will be used to do this over the next year. The data and information is being collected from the plans and plugged into the model. The results will be vetted through the subbasin planning groups and fish and wildlife managers. In addition, trend monitoring and evaluation needs will be identified and integrated with overall implementation of the program. This task will be completed by a regional group created to develop the protocols and monitoring needs for the basins. With subbasin plans now formally part of the fish and wildlife program, integration of monitoring and evaluation, and provincial goals, as well as project implementation consistent with the plans, has begun and will continue over the next several years.

"The plans helped to build better relations and deeper understanding with the public about complex and controversial environmental issues."

American Fisheries Society Symposium 52:111–115, 2006

Gearing up for Improved Collaboration: The Potentials and Limits of Cooperative Research for Incorporating Fishermen's Knowledge

Bonnie J. McCay, Teresa R. Johnson, Kevin St. Martin, and Doug Wilson

Department of Human Ecology, Cook College, Rutgers the State University
Cook Office Building, Room 210
55 Dudley Road, New Brunswick, New Jersey 08901, USA

Cooperative research provides a mechanism to renew trust and good faith in the management process, and contributes a sound methodological tool. In addition, it recognizes the expertise of different stakeholders. Cooperative research efforts also open the research process to greater scrutiny and increased transparency of the entire research process. Such programs can provide the key to improved relationships among marine stakeholders as well as more effective marine policy that is based on improved research design methodologies and appropriate data collection techniques—in other words, "the best science available." [Kaplan and McCay 2004]

Is cooperative research (CR) a way that fishermen's knowledge can be meaningfully integrated with scientific knowledge to improve fisheries science and management? Does CR improve conditions for mutual understanding and trust? Our research suggests that the answer is yes, qualified by the conclusion that it depends on the kind of cooperative research, the kind of knowledge, and the kinds of science and management questions.[1] Another condition is who actually participates.

Of particular importance, we argue, is developing mechanisms for the review of knowledge that comes from fishermen and out of cooperative research. Both peer review and relevancy review are critical to the actual use of such knowledge. We know from our own experiences, as well as from a large literature in the sociology of science[2], that scientists, like all good professionals, create boundaries around their work—boundaries that in the case of agency science are also influenced by legislative mandates.

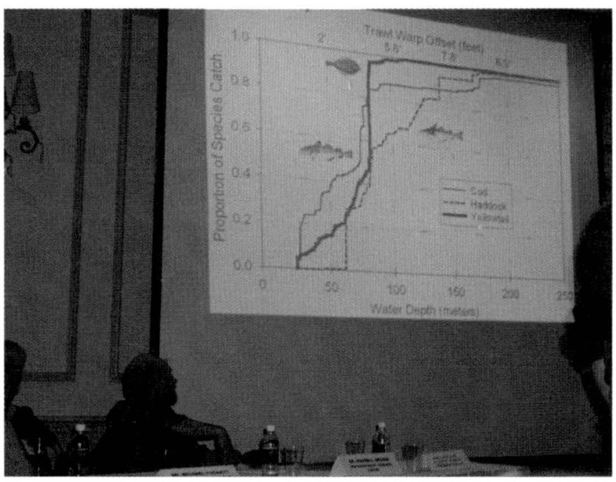

Photo 1. Steven Murawski, National Oceanic and Atmospheric Administration (NOAA)/Northeast Fisheries Science Center, reporting on the survey trawl problem at the Maine Fishermen's Forum, February 2003, explaining the problem of trawl warp offset in the NOAA trawl survey as part of cooperative work with members of the fishing industry to fix and improve the fishery-independent survey gear. Photo by B. McCay.

Cooperative research challenges the boundary-maintenance work of scientists, whose responses are important in determining what happens next.

Preliminary Research Findings

Here are some of the preliminary findings of our research to date, mainly based on interviews conducted by Teresa Johnson and Kevin St. Martin with people in the New England and mid-Atlantic fisheries and fisheries sciences:

1. There are three general kinds of fishery-oriented cooperative research: gear studies, stock assessment-related research, and ecological research.[3] Fishermen's experience-based knowledge (FEBK) can play a role in all three kinds.
2. Despite a history of limiting interaction to the use of the fishermen's vessel as a research platform, CR in the northeast has reached the point where it is expected that fishermen will have input to everything from project design to interpretation of the data. The implementation of these expectations is another story.
3. The fishermen and scientists involved greatly value the opportunity CR gives for improving their relationships, including changing attitudes and the ability to communicate, and this can take precedence over scientific outcomes.
4. *The fishermen involved appear to prefer gear studies most of all, and the scientists and technicians involved seem to feel that is the kind of cooperative research that is most useful.*[4] In gear studies, the experience-based knowledge of fishermen is more likely to be integrated with that of scientists and to be valued by scientists and managers than in collaborative research studies oriented toward stock assessment or ecological questions.
5. Consequently, the experience-based knowledge contributed by fishermen is more likely of the technical and operational type, rather than biological or ecological. *CR has not yet become a good way for the integration of fishermen's biological and ecological expertise, although there are exceptions and high hopes.*
6. More meaningful involvement of fishermen in stock assessment-related research is slowly taking place. One example may be the industry based surveys (IBSs) for Atlantic cod *Gadus morhua* and yellowtail flounder *Limanda ferruginea*, where half of the fixed sampling stations were selected by commercial fishermen.[5] Another such example is the Trawl Survey Advisory Panel, which includes fishermen and scientists working to improve the Northeast Fisheries Science Center bottom trawl survey.
7. In the northeast, very little FEBK used in cooperative research is ecological or environmental beyond not-at-all-trivial knowledge of how and where to find fish. Some of the projects that involve fishermen in environmental monitoring appear to be primarily "for-hire" type collaborations, where fishermen's contribution is primarily their vessels.
8. *Communication is a critical aspect of the working relationships central to cooperative research.* One of the reasons that gear CR seems to work so well is that the fishermen and the technical experts have training and experience in fishing gear and technology that enables them to "talk the same language."
9. In stock assessment CR aimed at improving fishery-independent or fishery-dependent data, the success of the CR effort sometimes depends on *the participation of academic scientists*, who are able to communicate with government scientists about technical matters concerning stock assessments, but who are also better able to communicate with fishermen, in part because they are not associated with the regulatory agency.
10. The issue of how cooperative research findings are used in fisheries management decision making is very challenging to all involved, given Data Quality Act, "best available science," and other legal mandates. Institutional factors such as legal mandates can continue to pose challenges to effective collaboration.[6]

11. Other challenges include unanticipated delays and a learning curve in getting many of the CR projects completed in ways that meet basic scientific requirements. Peer review is being handled within the National Marine Fisheries Service (NMFS). The New England Fishery Management Council and the Northeast Consortium are also engaged in implementing a process of relevancy review.
12. Culture is not necessarily the barrier between fishermen and scientists. Both fishermen and scientists often rely on anecdotal information to come to conclusions, and both recognize the value of objectivity and careful observation, although they usually use different tools and have different objectives. Cooperative research brings the objectives closer together and also can result in shared use of at least some aspects of scientific methods.
13. As CR develops, some scientists and fishermen have become very experienced and better prepared for it, improving the chances that it will be useful in the management process.
14. At the same time, though, participation in CR remains very narrow and may become more selective, reducing the value of CR in improving communication and trust between scientists, managers, and members of fishing communities.

Boundary Issues and Gear Studies

We conclude this brief essay with thoughts and comments on gear studies and boundary issues.

Gear studies do not really challenge traditional practices in the production of scientific knowledge. They especially do not challenge the authority of fisheries science done by NMFS scientists. This is probably because gear studies are not often done routinely by NMFS scientists anyway. For example, the infamous "trawl gate" could be blamed on the lack of expertise on the research vessel about gear technology and vessel operations.

In addition, gear studies have a long history and really cannot be done without industry boats. Examples of older industry-science collaborative gear studies include the Nordmore grate in the shrimp fishery and various studies to reduce the take of marine mammals. Many of the "collaborative" studies in the past did not fully involve fishermen in all phases of the research program, treating them mainly as owners and operators of "for-hire" vessels. Gear scientists doing collaborative research today emphasize

Photo 2. Rutgers University and National Fisheries Institute conduct collaborative gear selectivity research to reduce discards of scup *Stenotomus chrysops* in the longfin inshore squid *Loligo pealeii* fishery, Photo: Sarah King, Rutgers University, Haskin Shellfish Research Laboratory.

the difference. So in this way, the boundary between fishermen and science has been opened slightly in that the fishermen involved are now active participants in all aspects of the research projects.

Despite the fact that gear studies have been around a while and do actively involve fishermen and their knowledge, science still has the final say as to whether the results of the effort are determined to be of sufficient quality to be considered by management. The results of collaborative gear studies are sent for peer review just like other scientific projects. Science is viewed as a necessary instrument with which to gauge the validity of the outcome of gear studies. When asked what scientists bring to these collaborative gear studies, one participating independent scientist emphasized the need for scientific credibility in these studies to make them acceptable to management.

Many gear studies have been done and not used in management.[7] However, ideally, collaborative gear studies function to translate FEBK into something useable in the science policy process. In some cases, these gear studies are used to "prove" what fishermen already know about existing gear—either that it does or does not catch a certain fish species—or that fishing can occur with traditional or adjusted gear during certain times of the year in specific places with minimal bycatch of species of concern. Again, FEBK must be "proven" or validated by science in order to be useful for science or management. Otherwise, it is simply dismissed as anecdotal or as not widely applicable.

By and large, gear studies do not produce results that could be used to challenge stock assessments done by NMFS scientists and are therefore less likely to elicit boundary-maintenance responses. Gear studies do not collect information that would clarify total biomass removals (landings and discards), relative abundances, or life history characteristics of species. These are the kinds of information used in the stock assessment process. The objectives, methods of data collection, and types of data collected by gear studies are fundamentally different than what is needed to produce stock assessments. The stock assessment process is where federal fisheries science interacts most with fisheries management. Stock assessment advice really sets the grounds for what fisheries managers can do. Results of gear studies simply provide tools to meet management goals determined by stock assessment advice.

"...the boundary between fishermen and science has been opened slightly in that the fishermen involved are now active participants in all aspects of the research projects."

References

Kaplan, I. M., and B. J. McCay. 2004. Cooperative research, co-management and the social dimension of fisheries science and management. Marine Policy 28(3):257–258.

Wilson, D. C., and P. Degnbol. 2002. The effects of legal mandates on fisheries science deliberations: the case of Atlantic bluefish in the United States. Fisheries Research 58:1–14.

Footnotes

[1] These remarks are based on an ongoing research project, funded by the National Science Foundation, on the relationships between "experience-based" and "research-based" knowledge in fisheries science and policy: 2004–2007, Experience based knowledge in a science policy context, National Science Foundation, NSF 01-152, Societal Dimensions of Engineering, Science, and Technology (SDEST), Award no. SES-0349907. March 15, 2004–February 28, 2007. PI: Bonnie J. McCay; Co-PI: Kevin St. Martin; Graduate Research Assistant: Teresa Johnson. Douglas Wilson of the Institute for Fisheries Management in Hirtshals, Denmark is a consultant to the project, which involves comparisons with similar matters in Europe. Our study is not entirely about cooperative research, but cooperative research quickly emerged as the place where these relationships come to the fore, and thus, it has turned out to be the major focus.

[2] A large literature in the sociology of science is relevant. References to literature on the sociology of science are available from the lead author.

[3] For a good description of the cooperative research programs from a NOAA perspective, see "Overview of Regional Cooperative Research Programs, April 2005," http://www.st.nmfs.gov/st4/documents/cr/general_program_summary_2005.pdf. Also, see the Cooperative Research Partners Initiative Web site of NOAA, http://coopresearch.nero.noaa.gov/study.htm. Useful information on specific projects is available for the mid-Atlantic region at the Mid-Atlantic Fishery Management Council's Web site (www.mafmc.org) as "2002-2005 RSA Project Status as of 2004-10-01," http://www.mafmc.org/mid-atlantic/rsa/2002-2005_RSA_Project_Status_as_of_2004-10-01.pdf. http://www.northeastconsortium.org is the Web site for the Northeast Consortium, which manages many of the cooperative research projects of New England. www.fishresearch.org is a northeast Web site intended to be interactive between fishermen and scientists but apparently has not been updated since 2003.

[4] Gear studies seem to be the most popular of the collaborative research projects. The majority of the collaborative research projects funded through the Northeast Consortium and NOAA's Cooperative Research Partners Initiative program have been gear technology studies. In fact, one of the issues often raised about cooperative research in the region is that there are not enough gear technicians or experts available to work with all of the fishermen interested in doing collaborative gear work.

Why do fishermen and scientists collaborate so much on gear studies? One reason is that fishermen do contribute their expertise to these projects. A scientist explained that fishermen are a natural fit for gear studies because they are inherently testing gear as part of their livelihood. As is true with other types of collaborative research, fishermen who participate in gear studies receive financial benefits. The boats used in these studies are chartered. With increasingly restrictive regulations, fishermen are fishing fewer days a year. Collaborative gear studies keep their boat active for more of the year. These studies provide significant supplemental income to fishermen when they are not able to go fishing. There is another side to this, however. In some instances, participation in collaborative research may result in the loss of fishing revenue—this has to be balanced. Another example is that now fishermen in New England are being forced to utilize their individual "A" Days-at-Sea (DAS) allocations, which are more valuable than their "B" DAS because of concerns about additional mortality caused by research.

Fishermen also support gear studies because they have obvious benefits to management, as described above. These studies are very results oriented, and fishermen will likely see results more quickly compared to other studies that require long-term time series of data. Gear studies provide potential immediate opportunities for fishermen to harvest fish stocks that are otherwise closed to them because of bycatch concerns. For example, collaborative gear work provided fishermen with a new fishing opportunity in the whiting fishery, which otherwise would have been closed to them due to the need to rebuild groundfish stocks.

[5] In the Cooperative Research Partners Initiative cod and yellowtail IBSs, the industry selected about half of the stations, which are considered fixed. In some instances, the captains of the IBS are given flexibility in towing in a defined area around what is otherwise a station's randomly chosen location. One of the criticisms of NMFS surveys is that because of the random stratified design of the survey, the research vessel often tows in places where fishermen know that there are no fish. Allowing fishermen to contribute by selecting fixed stations or where to tow once in a place is important to addressing this concern of the industry.

[6] In an earlier project funded by the National Science Foundation, Science and Citizen Participation in Fisheries Management (Award No. 9810100, Bonnie J. McCay PI; Douglas Wilson co-PI), which ran from 07/1998 through 02/2001, we found the following: legal mandates about scientific issues affect the use of scientific data and claims about credibility. We found this to be true to a degree that challenged a common assumption in the literature that breakdowns in communication caused by a gap between training and experience, expressed as differing "world views" or knowledge cultures, were the primary obstacles to making use of experience-based knowledge (EBK) in fisheries management. Although these differences exist, the rules and institutions appeared to be the major blocks. This was particularly true in the case of bluefish *Pomatomus saltatrix* management where the use of EBK was blocked despite the fact that many of the scientists involved actually agreed with the substance of the EBK. The emphasis on processes in the present proposal is built on this finding and the need to explore its implications in other contexts. See Wilson and Degnbol 2002.

[7] Examples we have found thus far of northeast gear studies used in management include Nordmore grate research, research for the raised footrope whiting fishery, gear work concerning access to a closed area for yellowtail flounder, research to allow haddock *Melanogrammus aeglefinus* hook fishing as a "special access program" (SAP), and possibly a haddock separator trawl study.

Lessons Learned

American Fisheries Society Symposium 52:119–122, 2006

Lessons Learned: Making Cooperative Research and Cooperative Management Work

RAY HILBORN
University of Washington, School of Aquatic and Fishery Science
Box 355020, Seattle, Washington 98195-5020, USA

My participation in this symposium was invited, I suspect, because I was the chairman of a U.S. National Research Council (NRC) panel on "cooperative research in the National Marine Fisheries Service" that resulted in an NRC report of the same name. Much of the comments I make below are a result of the testimony presented to this panel and discussions with a broadly experience group of panel members, who included Joe DeAlteris, Rick Deriso, Ellen Pikitch, Mark Lundsten, Kees Zwanenburg, Gary Graham, Suzanne Iudicello-Martley, Gil Silvia, Priscilla Weeks, and John Williamson. Terry Schaefer was the NRC staff member who organized the panel. In the discussion below, I will provide some of the recommendations that emerged from the panel report, but otherwise, the comments reflect my own personal opinions and not those of the panel.

My own personal experience in cooperative research has been in a range of fisheries, primarily in New Zealand, western Canada, and Alaskan salmon, so most of this has not been in programs explicitly working with the National Marine Fisheries Service (NMFS), and in my comments here, I wish to concentrate on the more general questions of making cooperative research work, rather than the specifics of cooperative research with NMFS.

"The fishing industry possesses a wide range of knowledge about the dynamics of fish populations—they are on the water far more than any government agency could possibly be—and they possess many of the skills necessary for data collection and research."

The NRC panel concluded that cooperative research is the natural way to conduct a wide range of fisheries research activities and should be the "default option" when new data collection programs are considered. This is certainly not the current culture in most national and state fisheries agencies. Many years ago, I attended a workshop where the keynote speaker was a commercial trawl fisherman from Oregon who began his talk by saying "my job [the commercial fishermen] is to catch as many fish as possible, your job [scientists and managers] is to keep me from catching too many." This quote is a reflection of the prevailing culture of the time; agencies collect data and manage, while fishermen catch fish.

The fishing industry possesses a wide range of knowledge about the dynamics of fish populations—they are on the water far more than any government agency could possibly be—and they possess many of the skills necessary for data collection and research. Furthermore, the industry itself has the infrastructure to sample the fishery, and in many cases the fish populations, far more intensively and reliably than government agencies. To exclude this knowledge and infrastructure would be a very serious mistake.

Considering cooperative research as an umbrella term for activities that involve both government and stakeholders, there is a continuum of levels of cooperation that may range from long-term survey programs conducted almost exclusively by government, perhaps with some review by stakeholders, to data collection programs or research programs conducted almost exclusively by stakeholders with some review by government. An example of the later is the seabird bycatch reduction study by the North Pacific longline fleet, Sea Grant, and university scientists.

Cooperative research may be one-off, stand-alone projects, such as the seabird bycatch study mentioned above, or it may be long-term projects that form the core of a fisheries management program, such as the New Zealand rock lobster industry's length frequency sampling program or the Canadian Sablefish Association's survey and tagging program.

If we wish to understand why cooperative research happens in some places and times and does not in others, we need to look primarily at the incentives, both for stakeholders and for government agencies. Most of the cooperative research programs have involved commercial fishing groups, but there are certainly good examples of native groups, recreational fishermen, and conservation NGO's working cooperatively with governments as well. Commercial fish groups get involved in cooperative research when they see a direct benefit to their own interest. Thus, you see the largest and most sustained cooperative research programs in fisheries where there is some form of long-term dedicated access, the individual transferable quotas programs in New Zealand, Canadian sablefish *Anoplopoma fimbria*, and the U.S. pacific halibut *Hippoglossoides stenolepis* and sablefish being good examples, the Indian tribes of the Pacific Northwest another. In these fisheries, the participants have strong financial incentives to make management better and very often put up funds to pay to have the research done.

I would contrast those types of fisheries, dedicated access equals financial incentives, to the more common situation in Canada and the United States, where the primary motivation for cooperation has either been earmarked government funding of cooperative activities, as has happened in New England, or small-scale problem-solving projects that benefit a particular group of stakeholders, such as the scallop surveys conducted by industry in New England. I would note that there were certainly strong incentives for the New England scallop industry to conduct these surveys, and they do have a form of dedicated access through the licensing system.

In summary, if you want to understand why cooperative research happens, look at the incentives.

Making cooperative research successful requires more than incentives; it requires treating cooperative research in the same way as any other form of research, with proper scientific design, oversight, review, and publication. While stakeholders do have many strengths, they usually need to obtain external scientific input to help them design data collection and research projects. While commercial fishermen know how to catch fish and run boats, few understand scientific sampling. All of the successful cooperative research programs we considered involved significant participation by scientists, either from government agencies or externally. The New Zealand Seafood Industry Council, which is an umbrella group involving all commercial fishing interests in New Zealand, has a staff of about 10 scientists who work on data collection, survey design, and all aspects of the science programs conducted by various New Zealand seafood industries.

"If we wish to understand why cooperative research happens in some places and times and does not in others, we need to look primarily at the incentives, both for stakeholders and for government agencies."

The specific discussion questions posed are given below, and I make comments relevant to each.

What Are the Characteristics of Successful and Unsuccessful Cooperative Research Projects? How Do They Differ on Each Coast, by Scale?

Here the answer is simple, incentives, and organization. The reasons for programs "failing" were often lack of incentives on both sides—for instance, lack of incentives for individual NMFS scientists to participate in such programs, the subject of a specific recommendation of the NRC panel. Organizational failures included the mundane, permits, and insurance (again subjects of specific NRC panel recommendations) or more broadly included poor scientific, organizational, or fiscal oversight.

What Kind of Leadership Is Necessary?

This question is difficult to answer simply; almost all projects had someone who took charge and led the program. Again, there was usually a strong incentive for this person to do so.

How Is Local Knowledge Integrated into the Scientific process? What Roles Are All Parties Playing throughout the Scientific Process?

There is no generic answer to this; almost any project requires local knowledge, perhaps in how to rig gear on a specific type of boat, perhaps in understanding the bottom in designing a survey. Often, stakeholders have a broad range of such knowledge.

What Data Quality Activities Work (e.g., Peer-Review, Data Management)?

It is particularly important for cooperative research to have intensive peer review and monitoring, primarily because its credibility needs to be established, but more importantly because science used for public policy decisions needs to be thoroughly reviewed.

The Recommendations of the NRC Panel

- Cooperative research should be considered as a usual and normal approach for conducting fisheries research.
- Congress and NMFS should give serious consideration to establishing and funding regional research boards to prioritize and coordinate the use of dedicated funding (earmarks and line items) for cooperative research projects in each region; evaluate NMFS dedicated research projects for its potential as cooperative research; foster communication of research results, and evaluate cooperative research projects and programs.
- In designing cooperative research projects, the applicability of the results to the overall success of the fishery must be considered.
- A majority of cooperative research funds should be allocated through a competitive review process. The remaining monies should be used for rapid response, seed money, and administration.
- Expectations, requirements, and procedures, including the development of agreements carefully detailing the responsibilities of all participants, should be clarified at the beginning of every project.
- Commercial fishing vessels used for cooperative fisheries research by NMFS should meet all U.S. Coast Guard requirements for operation and manning so as to ensure safe operations.
- NMFS should ensure that appropriate liability insurance is secured for all cooperative fisheries research activities so as to protect the financial interests of all participants involved in cooperative research.
- NMFS should streamline and standardize all permitting procedures for conducting cooperative research projects so as to ensure uniform treatment and rapid processing of all applications in all regions.
- NMFS and operators of commercial fishing vessels should use comprehensive contracting procedures so as to minimize confusion and maximize opportunity for all fishermen to participate in cooperative research.

- All participants in cooperative research should ensure the independent status of observers and the confidentiality of the data collected.
- Appropriate administrative overhead should be included in all budgetary allocations.
- NMFS should recognize and hire individuals with the interpersonal and communication skills necessary for cooperative research.
- NMFS should progress reports, and dissemination research project plan.
- NMFS scientists that participate in cooperative research should be given equal opportunity for professional advancement *along with their counterparts that do not participate in cooperative research.*

American Fisheries Society Symposium 52:123–128, 2006

Session 2A: Lessons Learned

Making Cooperative Fisheries Research Work, Panel Summary

Roland Barnaby
New Hampshire Sea Grant Extension
113 North Road, Brentwood, New Hampshire 03833, USA

> The core of the program is the strong partnerships. A strong relationship allows partners to move forward and complete tasks. [Tom Rudolph, research director, Cape Cod Commercial Hook Fishermen's Association]

The second session of the joint Sea Grant/American Fisheries Society symposium focused on the lessons learned from cooperative fisheries research and management. Rolland Barnaby, with the New Hampshire Sea Grant Program, moderated the panel. Below are some speaker highlights, followed by a summary of the facilitated panel-audience discussion. White papers prepared by the keynote and panelists follow this panel summary.

As chair of the recent (2004) National Research Council (NRC) review of cooperative research in the National Marine Fisheries Service (NMFS), Ray Hilborn set the stage for this session, defining the broad range of cooperative fisheries research and management and outlining the range of research and management options.

Ray Hilborn, professor of fisheries management at the School of Aquatic & Fishery Sciences, University of Washington

- The dominant paradigm in marine conservation is top-down—you have an agency that collects the data, analyzes it, and makes decisions, and funding comes from legislatures. One concern with this model is the limited science training prevalent among the congressional staffers that are often making significant marine policy decisions.
- Community-based conservation, in the extreme, is complete devolution of authority to communi ties; resources are owned and protected and managed by the community.
- With cooperative management, you are basically sliding between top-down control to community-based control.
- When regulations spread through the industry voluntarily, it works better. Fishermen have experience and understanding; how can we take advantage and incorporate this knowledge?
- Dedicated access programs (e.g., individual transferable quotas) can generate economic profitability and incentives to improve research programs; stakeholders respond to threats to their profit and their resource. These allocation approaches cause groups to take independent action when there is a threat to their profit, and this becomes the incentive to improve research.

- We need more data because the bar has been raised. The level of accuracy demanded by current standards in fisheries research and management is much higher than in the past. Employing cooperative fisheries research is one way to start gathering more data in a more cost-effective way.

Summary and Overview of Panel-Audience Discussion

At the completion of the presentations, the moderator facilitated a discussion between panelists and audience members to address the lessons learned from cooperative fisheries research, including characteristics of successful projects, leadership, challenges, knowledge integration, program monitoring, and the future of cooperative fisheries research.

Characteristics of Successful Cooperative Research

Several speakers indicated that successful cooperative research projects needed buy-in from affected parties at the beginning of the process. Gordon Curry gave two examples of successful cooperative research projects, but one of them had trouble utilizing the results in management. The project that had better integration had buy-in from the industry at the beginning of the project. Curry said their active project teams, composed of diverse stakeholders, were a key part of the process. A project team needs a knowledgeable fisherman, a fisheries manager who can see the big picture, a scientific author-

Elizabeth Clarke, National Oceanic and Atmospheric Administration, division director, Fishery Resource Analysis and Monitoring Division, Northwest Fisheries Science Center (substitute for John Watson, Southeast Fisheries Science Center in Pascagoula, Mississippi)

- Continuum from cooperation to collaboration; the cooperative fisheries research program really seeks to have projects in the full continuum. Wants to design a project to provide for full continuum involvement in the process of cooperation to collaboration.
- Lessons learned. Not everyone participates in the same way. You need a diverse program if you want the widest possible participation. All types of cooperation in the continuum to collaboration have improved our science and improve industry support of National Marine Fisheries Service science.
- Mantra. "Promise less, deliver more." Do not over promise; it is better to work methodically and promise what you can deliver and then maybe more.

Tom Rudolph, Cape Cod Commercial Hook Fishermen's Association, Massachusetts

- The core of the program is the strong partnerships. A strong relationship allows partners to move forward and complete tasks. The program is split into four broad areas: marine education, policy, community-based management, and cooperative fisheries research.
- The mission of the Cape Cod Commercial Hook Fishermen's Association is to advocate for a healthy marine environment and sustainable fisheries. Protect both fish and fishermen. Focus is on achieving sustainable fisheries in a number of ways. One way is to advocate and empower hook and line fishermen to get involved in the cooperative process in order to preserve their way of life and way of fishing.

ity, a representative from a nongovernmental organization (NGO), and the enforcement community if the research is going to impact regulations. Another level of team members can come from state and federal government, academia, decision makers, and the general public. This team framework was invaluable to success.

Clear and concise agreement on everyone's role in a cooperative research project is important. Elizabeth Clarke described an outreach program to tell everyone, including academics and NGOs,

- Under the Canadian Fisheries Reform policy, new conservation needs must be met, while considering how to maintain harvester viability.
- Tools need to be developed to meet conservation standards while maintaining viability. Harvesters submit proposals to create more responsible fisheries. Through a formalized collaborative agreement, project teams work together to complete projects, involving many partners pulling together to accomplish goals. Then results are peer reviewed. Partners work together to implement their findings.
- A clear policy with goals outlined, detailed project proposal to build from, and acceptance from involved groups (respect, trust) helps progression. Independent analysis and review is integral.

Gordon Curry, Department of Fisheries and Oceans, Canada

Mike Matyelwich, Columbia River Inter-Tribal Fish Commission

- The Columbia River Inter-Tribal Fish Commission works with tribes to manage and protect treaty-reserved fisheries resources.
- The legal framework established cooperation and forced the interaction between decision makers and scientists. It helped with conflicts, created structure to resolve disputes, and ensured accountability among partners.
- Communication allows people to work together. They have developed common research criteria, which forces consensus building and creates useable products for decision making. Legal framework allows for accountability and forces people to look at objectives and what data are needed to make decisions and develop evaluation criteria.
- Leadership. A leader needs vision, empathy, curiosity, and perspective. He/she needs a sense of character. He/she needs a sense of the larger purpose and needs to have a mental model of how things work. An ability to deal with change, rather than just complexity.
- Tribal decision makers usually have been fishermen. Hence, they have first-hand knowledge of the fish species, their habitats, and their behavior that the state managers might not have. Local knowledge is important; it should be shared and not lost.

about the proposed program that she has used and that has been helpful. This has increased awareness and support of the program. She also suggested that participants are able to add value to the project in different ways; a diverse program with a variety of avenues to participate will be successful, allowing for the broadest range of participation.

Motivations for involvement in cooperative programs vary. The NRC panel evaluated the motivations for individuals involved in cooperative programs. Fishermen seem to participate when either they needed to avoid a problem or increase revenue. For National Oceanic and Atmospheric Administration (NOAA) staff, motivation is difficult to evaluate because the internal incentive structure within the organization is different. If they have a supportive superior, then they will participate; if they do not, then they will not or cannot participate. The NRC panel made recommendations about providing more incentives to be involved in cooperative research within NOAA. It is necessary to have an agency leader to support volunteering in cooperative programs. Often times, the motivation for the research comes from declining fish stocks and industry questioning the quality of science and the validity of NOAA/NMFS science programs. It can be a challenge getting the right people at the table and buy-in from all the stakeholders at the very beginning of the project, but a crisis or perceived crisis gets people involved.

Leadership

Leadership for most of the cooperative fisheries research and management projects came from a group of stakeholders from different geographic regions or a political agency with stake or interest in a shared common resource. Where there is a community-based management structure, the community often becomes the leader in promoting and developing cooperative research projects.

Michael Matylewich suggested that a leader needs a vision, a sense of the larger purpose and the ability to deal with change, rather than just complexity. In most of the examples presented in the session, a leader was necessary to get the involved parties together to start a project. Organizing and coordinating mechanisms, clear roles and responsibilities with a wide range of options for participation, capable partners participating, and independent analyses of methods, results, and findings were all mentioned as necessary elements for successful leaders to promote.

Several speakers mentioned trust and respect as important factors in a successful project; trust and respect were also byproducts of successful projects. What types of factors are important in generating trust and making cooperation work? Personalities are important, picking the right collaborative partners where they are able to leave personal agendas at the door and are willing to work toward a common goal.

Challenges

Funding is always a challenge but may be easier to obtain with a broad group of stakeholders working together. One way to address funding needs is to provide opportunities for volunteers and to account for in-kind contributions or direct research funding from industry in other funding proposals, which demonstrate leveraging resources efficiently.

The biggest challenge for cooperative research is the process of integrating results into the management arena. Several projects indicated problems in this area. There are many affected parties involved, and therefore, people need to work together on building consensus. Trying to align all of the different interests can be very difficult; it takes open and uninhibited communication, commitment to the goals, as well as commitment of resources (both time and money) on the part of each group involved.

Panelists discussed how to better organize partners. It was stressed that there needs to be encouragement for more involvement. The bottom line in most cases comes down to developing and building professional working relationships. The groups need to work on their relationship, and it needs to be built on trust. Resources and tools used need to be creative. Strong, well-organized fishing organizations can be a great help in the development of cooperative research programs because the necessary buy-in can come from within and can be promoted by the organizations. Tom Rudolph from the Cape Cod Commercial Fishermen's Association indicated that the core of their cooperative research programs is based on strong partnerships. In their sector, fisheries management plans and harvest agreements are signed by the organization, the organization has a legally binding agreement with the federal government, and this works well.

Knowledge Integration

Getting local knowledge integrated into the management process was difficult in many of the cooperative research projects. It was easier when there was a strong partnership between the fishing industry, the management agency, and NGOs. The integration of knowledge between stakeholders may be one of the most significant outcomes from successful cooperative fisheries research and management projects. Panelists talked about the true benefits that are the product of the process by which fishermen, managers, scientists, and NGOs exchange information and appreciate and use one another's expertise.

The devolution of authority produces an environment that enhances the perceived and real need for cooperative fisheries research and knowledge integration. Ray Hilborn spoke about cooperative fisheries research and management moving from top-down conservation to bottom-up control. Joint efforts to collect and analyze data create better-informed fishermen who tend to use their expertise in the management process more effectively. The panel discussed strategies for gaining industry support for collaborative projects, including the establishment of a thorough consultation process that prevents personal agendas from inhibiting progress, clearly defines project goals, and promotes data sharing. This will aid in the professional relationship building process. When there are negative perceptions, they will take time to overcome, but you need buy-in from the start; then, you can begin to overcome negative misconceptions before they become obstacles to progress. Devolution of authority allows groups to develop an infrastructure and manage themselves; this allows the government to be the facilitator rather than the sole responsible decision-making authority. Regulations developed by the industry are much more effective than ones solely developed by government. An example of where this has worked and has been a true collaboration is an acoustic survey that is conducted on a commercial vessel. The survey was designed from the ground up within the fishery. There have been periodic group meetings to talk about the design and implementation, and this has worked well.

Panelists shared that there are many fishermen willing to support and help in cooperative research, but it can be difficult for them; often their boats are not outfitted for research and data collection, it takes away from fishing days, and it is expensive to run the boat. One way to alleviate this cost for conservation engineering/gear research would be to allow the sale of fish caught during research projects. The challenge is to provide sufficient financial incentives for industry to cover expenses and promote their involvement, while not creating an incentive to fish. Fishing and research are distinct activities.

Monitoring Project Effectiveness

Elizabeth Clarke acknowledged that the NOAA Fisheries Northwest Fisheries Science Center has experienced impacts that are difficult to quantify from cooperative fisheries research. For example, it is hard to measure the value of being able to go to a public hearing and have a constructive, nonadversarial conversation. She shared that without the cooperation of industry vessels, often times the amount of work that is completed would have been otherwise impossible (e.g., West Coast groundfish surveys). Also, industry participation in the council process and in the review processes has changed the dynamic of the acceptance of science by industry. It has smoothed the decision-making process of how cooperative fisheries research results are moved into the management arena.

It was also stressed that independent review and analysis are essential. Common analyses and review processes are helpful. Gordon Curry also suggested that enforcement of the plan after implementation, monitoring for follow-through within the plan, as well as closing the information loop are essential ingredients for success.

What's Next?

The number and diversity of cooperative research projects that were presented was impressive. All participants thought cooperative research and cooperative management was going to continue, and most indicated it was going to increase. As Hilborn pointed out in his keynote address, cooperative research "should be the standard method of operation in the future, it should be the norm"; there may be different methods of integration, but as long as the data are collected in a rigorous way, things are moving in a positive direction. As the number of cooperative research projects increases, so does the prevalence of cooperative management and vice versa; they encourage and enable each other. This bodes well for both.

American Fisheries Society Symposium 52:129–133, 2006

Making Cooperative Research Work: A Move to Collaborative Salmon Management and Research on the West Coast of British Columbia

GORDON R. CURRY
Fisheries and Oceans Canada
3225 Stephenson Point Road
Nanaimo, British Columbia V9T 1K3, Canada

Introduction

In Canada, wild Pacific salmon are managed as a common property resource through licence access and catch sharing between sectors through an allocation policy. Alternate management strategies are being considered and may include more defined shares and a greater sense of ownership, combined with increased responsibility by harvesters for management and research. Fisheries and Oceans Canada (DFO) funds most of the costs of management and research in the salmon fishery, but with shrinking government budgets, harvesters are being asked to take on a greater role, including funding management and research.

“In Canada, wild Pacific salmon are managed as a common property resource through licence access and catch sharing between sectors through an allocation policy.”

Concern for “species at risk” under the Species at Risk Act (SARA) and other stocks at low abundance are now driving the management and therefore access to abundant stocks (“weak stock management”). The current management regime and harvest strategies have resulted in a fishery struggling through change and having difficulty attaining viability. First Nations, Industry, other stakeholders, and DFO are working through a new process called “Fisheries Reform.” This program aims to explore ways to increase viability in the fishery, meet the ever increasing conservation standards while also working towards greater cooperative management. There are four main themes for Pacific fishery reform:

- Sustaining strong salmon populations by setting clear conservation objectives based on the principles of the Wild Salmon Policy;
- Strengthening DFO programs that are critical to conservation, such as habitat protection, enforcement, and the scientific assessment of stocks;
- Making progress over time on increasing First Nations’ access to economic fisheries in collaboration with First Nations and Indian and Northern Affairs; and
- Improving the economic performance of fisheries so that they reach their full potential, provide certainty to participants, and optimize harvest opportunities.

Catalyst for Change

The catalyst for the move to weak stock management was the collapse of many coho salmon *Oncorhynchus kisutch* stocks coast-wide (the “coho crisis”) in the mid- to late 1990s, which resulted in a government program to respond to the downturn in production. The Canadian Fisheries Adjustment and Restructuring Program provided funding to research selective fishing gear and methods. The Selective Fishing Program was funded from 1998 until March 31, 2002. Approximately 21.5 million dollars was spent on selective fishing research, implementation, training, and education to start a paradigm shift focusing on the protection of weak species and stocks. Additional funding came from the use of up to five percent of the coast-wide total allowable catch (TAC), which could also be used to fund commercial research projects. The focus of the Selective Fisheries Program was to find gear and methods that either avoided species or stocks of concern or if encountered allowed their release with the least harm.

The Selective Fisheries Program (1998–2002)

The objective of the Selective Fisheries Program was to increase the selectivity of harvesting practices and technologies to allow harvest of surplus stocks while protecting stocks at risk. This government funded program resulted in 63 research projects to investigate encounter rates, catch monitoring, catch mortality, stock identification, and other aspects. Many of the 122 additional projects included varying levels of scientific design to find selective fishing solutions to mixed species and stock fisheries. Harvesters were solicited to submit selective fishing proposals taking into consideration various criteria. These proposals were rated and ranked against predetermined criteria such as conservation, study design, budget, future applicability, and so forth. Initially, this ranking process included DFO and provincial fisheries staff but in subsequent years included increasing input from all sectors. Successful candidates were then organized to meet with a project team that often included the following:

- Proponent to present the proposal, maintain the practicality of the proposed work, manage the project, and carry out the work;
- Fisheries manager to facilitate and coordinate the process;
- Scientific authority to ensure the study design was appropriate and maintained for the questions posed;
- Independent technician to gather the field data; and
- Others participated as required, such as associates of the proponent, nongovernment organizations, DFO enforcement staff, or other technical and academic assistance from outside DFO (contract biologists and university scientists).

Fisheries Mangers Maintained the Focus on Selective Fishing (2002–present)

Since the Selective Fisheries Program ended in 2002, government fisheries managers were provided direction from the Selective Fisheries Policy and mandated with facilitating continued research on selective gear and methods. The commercial sector and the policy based use of 5% of the TAC continued to allow experimentation. Unfortunately, the First Nation and recreational sectors have no easy source of funding; therefore, research has been significantly reduced. Even with access to 5% of the TAC, some commercial licence area groups have been less than strategic in finding conservation-based solutions to maintain or increase their access to abundant stocks. Other groups with a strong representative organization have continued to make progress.

> “DFO is experiencing budget downsizing necessitating that those profiting from the harvest of fish fund a greater proportion of the management costs.”

The Selective Fisheries Policy also called for standards to be developed to clarify for all harvesters how high the conservation bar had been set. These standards have yet to become formalized; therefore, harvesters are unclear as to how selective their fisheries must become. The advent of the SARA and recent release of the Wild Salmon Policy (WSP) will again fuel the need for government to clarify conservation standards for fisheries, which in turn will guide where research should be focused.

Moving to Collaborative Agreements and Comanagement Funded Research

DFO is experiencing budget downsizing necessitating that those profiting from the harvest of fish fund a greater proportion of the management costs. Cost recovery has been part of many commercial fisheries for years but has only started to be implemented in the salmon fishery. Asking each licence holder to write a check to an industry organization for comanagement usually results in a negative response. It is easier for licence holders to accept the harvest of a portion of their TAC to fund these costs. With greater investment in the cost of research and management comes a greater role in decision making and therefore greater control over their fishery.

One issue with the salmon fishery is a lack of viability. This problem has been compounded in recent years by the switch to weak stock management. This has resulted in significant portions of available catch not being accessible due to an unacceptable conservation cost with its harvest. Allowing for creative harvesting of a portion of available TAC has resulted in some licence area groups catching sufficient fish to fund their annual costs of research and management.

“DFO has encouraged licence area groups to enter into formal collaborative agreements with government in order that they may take greater control of their fishery while funding a greater proportion of research and management.”

DFO has encouraged licence area groups to enter into formal collaborative agreements with government in order that they may take greater control of their fishery while funding a greater proportion of research and management. They outline the cooperative research and management projects and costs proposed for the coming year, and this total cost then dictates the number of fish to be caught. The fishery to fund the agreement is licensed by a scientific licence to enable this fishery outside regular fisheries, and since there is a known amount of fish being caught, it allows the group to put the catch up for bid prior to the fishery, allowing for a greater financial return.

The Current Climate for Making Collaborative Research Work

Since the Selective Fisheries Program, there has been a paradigm shift to a conservation based fishery, driven primarily by consideration of weak species or stock management. Recent legislation (SARA) and policy (WSP) are requiring higher and more transparent standards to clarify the parameters around access to abundant fish.

Those groups having representative organizations and collaborative agreements have entered into a developing comanagement relationship with DFO. These groups have taken on greater control over the management and research required in their fishery. The improved relationship results in strategic discussion followed by planning for implementing research required to answer questions. These questions, and resulting answers (through research), are often crucial to providing a level of conservation comfort that allows the department to consider a fishery it may not have otherwise. If there is uncertainty about the conservation outcome of a planned fishery, the department, by policy, will err on the side of conservation. This policy has resulted in many fisheries not being opened due to the suspected but unknown impacts. It is in the best interest of the licence area group to anticipate conservation concerns and do the research required to find gear and method solutions.

In the past, many of the interactions between DFO and harvesters have been fraught with an adversarial climate. Those groups who have entered into collaborative agreements with DFO can be characterized as having a very good working relationship based on trust and respect. Another important aspect is the funding to pursue solutions, which require the combination of a healthy relationship and money to be effective. These effective working relationships are resulting in more of a team approach. I believe this will result in finding increased solutions that can lead to a more conservation based and viable fishery.

The following are two case studies that highlight actual approaches to research:

Case study no. 1—seine bunt grids (to allow non-target catch to exit the net)

Seine bunt grids are panels that are laced into the end of a seine net. These panels have holes to allow small fish to swim through while retaining the target catch. In 1998, a seine harvester completed a project using rigid plastic grids and proved that small salmon will use these devices as an escape route out of the net. Another harvester took up the work on designing a flexible grid while scientifically testing salmon preference for color, grid location in the net, escape-hole shape, and size. In addition, he experimented with a variety of grid materials until, in 2004, deciding on the overall best grid combination. The current grid is a clear poly vinyl chloride (PVC) with dimensions of about 24 × 36 in, with a number of oblong-shaped openings. An important companion to the grids is a small mesh, knotless bunt of about 60 mm mesh size. This knotless web reduces the abrasion on the catch, therefore reducing scale loss, and the smaller mesh size has significantly reduced gilling over the conventional 70- and 100-mm knotted bunts. The overall test of this gear combination is the in-water self sorting of non-target salmon catch and reduced mortality on those salmon requiring sorting and release. In one area, the mortality rate associated with released salmon has been shown to be reduced from about 70% mortality to below 25%. To date, the results have been very compelling.

> "Those groups who have entered into collaborative agreements with DFO can be characterized as having a very good working relationship based on trust and respect."

Key aspects regarding cooperative grid research are as follows:

- Skipper (Paul Brajcich) doing the research each year has a long family history in the fishery.
- DFO has continued to support this research (project team approach), due to the promising conservation results, in spite of the area licence committee's reluctance to accept these results and see the project proceed.
- There are 167 licenses in this group with nearly half controlled by one company.
- Research has stalled in 2005 due to a lack of industry support.
- Peer review of each year's research is the next step to confirm results (Fall 2005).
- If peer review is positive, DFO will be discussing possible further research and an implementation plan with industry.
- Progress can be characterized as having to be constantly pushed by DFO in favor of a conservation mandate while often disagreeing with the elected representatives.

Case study no. 2—gulf trollers move to collaborative management and research

The Gulf Trollers Association (GTA) is a group of 150 troll vessel licence holders who are licenced to fish in the waters off the East Coast of Vancouver Island. This group has lost access to Chinook salmon *O. tshawytscha* and coho salmon *O. keta*, their main target species of the past, due to conservation and allocation issues. They now try to survive on about 10 d of fishing on each of sockeye salmon *O. nerka* and chum salmon and now up to a couple of weeks on pink salmon *O. gorbuscha*. The GTA are now working their fourth year under a collaborative agreement with DFO. They have caught a small portion of their sockeye salmon TAC (10,000) using a seine vessel to fund cooperative research and management. These fish would not have been caught within the narrow conservation window using troll gear; therefore, this was an easy sell to the membership. The GTA annually decide in concert with DFO on the research projects they will support and fund. Their funding also covers comanagement costs, other nonresearch projects, and supplemental funding support for some management costs by DFO.

Key aspects of the GTA approach to making cooperative research work are as follows:

- Fourth year under a collaborative agreement with a trusting and effective relationship with DFO.
- Increasing control, funding, and support over their priorities and strategic direction.
- Research (project team approach) leading to solid arguments to maintain fisheries, increase existing access, or open new fisheries.
- The GTA are challenging DFO on the need for policy changes to allow for a different but viable fishery while meeting conservation standards.
- Progress can be characterized with the GTA as truly cooperative and strategic.

Ten Suggested Steps to make Cooperative Research Work

1. A policy environment to guide management and research towards a vision of the fishery.
2. A detailed project proposal.
3. Affected harvester organization to be in favour of proposed research and potential results.
4. A project team to review, modify, and implement the research project with solid study design.
5. Funding source to cover the project, initial analysis, and report.
6. Independent project analysis or peer review (best) of this and other related studies.
7. Development of an implementation plan for the new gear or method.
8. Government and industry organization to develop enforcement strategies to deter infractions.
9. Government to enact conditions and legislation to cover new fishing gear or methods.
10. Adaptive monitoring of fisheries to identify ongoing research needs.

American Fisheries Society Symposium 52:135–139, 2006

Making Cooperative Research Work: The Columbia River Fisheries Management Experience

MIKE A. MATYLEWICH
Columbia River Inter-Tribal Fish Commission
729 NE Oregon Street, Suite 200, Portland, Oregon 97232, USA

Adult returns of salmon *Oncorhynchus* spp. and steelhead *O. mykiss* to the Columbia River basin in the predevelopment era are estimated at 11–16 million (NPPC 1986). In 1995, adult returns hit a low of 672,000 (ODFW and WDFW 2002). Recent returns are 2–3 million annually. Many factors contributed to the decline, including hydropower development, irrigation, agriculture, forestry, urbanization, and fishing. The challenge today is managing a resource subject to continued development pressures. In the last several decades, fisheries managers have incorporated advances in data gathering to provide policy makers with an improved basis for making decisions.

The fisheries management process in the Columbia River changed significantly in 1969 when a group of tribal members sued the state of Oregon in federal court over fishing regulations (*SoHappy v. Smith* 1969). The case, which was consolidated with *United States v. Oregon* 1969, established and quantified the treaty fishing right. The legal framework resulted in establishing the treaty tribes as comanagers of the resource. The first fish management plan under court jurisdiction was completed in 1977. The focus of the 1977 plan, also referred to as the Five-Year Plan, was allocation of in-river fishing. The 1977 plan also created the Technical Advisory Committee (TAC) as a forum for the parties to develop and analyze biological information. The parties agreed to "diligently pursue and promote, through cooperative efforts, upriver maintenance of fish habitat and hatchery rearing programs," but did not include specific actions. When the 1977 plan expired in 1982, the *United States v. Oregon* parties began negotiating a successor plan.

Significant new information on the ocean distribution of salmon became available with the implementation of the coast-wide coded-wire-tag (cwt) program in the late 1970s. The system required the cooperation of agencies all along the Pacific coast to provide release and recovery data. The information allowed fisheries managers to quantify the contribution of tagged groups to ocean fisheries. The allocation discussions expanded to include the effects of ocean fisheries. The tribes sued the Secretary of Commerce each year from 1979 to 1982 (*Yakima v. Baldridge*1985) over ocean fishing regulations that did not provide for adequate inside/outside allocation.

In 1980, Congress passed the Northwest Power Act in response to potential listings of salmon under the Endangered Species Act (ESA) and created the Northwest Power Planning Council (NPPC) consisting of members representing Oregon, Washington, Idaho, and Montana. The task of the NPPC is to balance the needs of fish and wildlife with the region's need for electricity. The NPPC provided a forum for the region to consider the environmental factors leading to the decline of the salmon resource in the Columbia River. One of the early members considered the NPPC to be a regional experiment in adaptive management (Lee 1993). Local knowledge was incorporated into the process through subbasin planning. The intent of the *United States v. Oregon* parties was to utilize proposed actions developed through the subbasin planning process in the early 1980s to help achieve the rebuilding objectives of a new management plan. However, this intention was never realized because the instructions to subbasin planning groups were not explicit enough

"...diligently pursue and promote, through cooperative efforts, upriver maintenance of fish habitat and hatchery rearing programs."

and the resulting plans were not readily transferable to the *United States v. Oregon* process. A second round of subbasin planning is underway to identify proposed actions. The planning process will be used to help prioritize funding for restoration projects.

It took the *United States v. Oregon* parties 5 years to negotiate a new management plan. The Columbia River Fish Management Plan (CRFMP) was entered into court in 1988. The preamble to the CRFMP states that "the primary goals of the Parties are to rebuild weak runs to full productivity and fairly share the harvest of upper river runs between treaty Indian and non-Indian fisheries in the ocean and Columbia River Basin." The preamble also states the "the Parties intend to use (as herein specified) habitat protection authorities, enhancement efforts, and artificial production techniques as well as harvest to ensure that Columbia River fish runs continue to provide a broad range of benefits in perpetuity." The CRFMP had a duration of 10 years. The CRFMP was an improvement over the 1977 plan for several reasons. Allocation considerations were expanded to include ocean fisheries, based on information generated through the cwt program. Specific hatchery actions intended to assist in rebuilding were identified and committed to. The Production Advisory Committee (PAC) was created to develop and analyze information related to hatchery actions. The CRFMP also outlines a dispute resolution process for technical and policy issues.

At the same time that the *United States v. Oregon* parties were negotiating the CRFMP, the United States was negotiating the Pacific Salmon Interception Treaty with Canada. Discussions between the countries had gone on for several decades. One of the factors that changed the dynamics of the discussions in the 1980s was the additional harvest information provided by the cwt program. Settlement of litigation concerning the accounting of Alaska catch in treaty/nontreaty allocation was also a factor. The Pacific Salmon Interception Treaty was completed in 1988. The Pacific Salmon Commission (PSC) serves as the coordinating agency for implementation of the treaty. One of the goals of the PSC was to achieve coast-wide rebuilding of Chinook salmon *Oncorhynchus tshawytscha* stocks by 1998. A number of technical committees were established to promote the development and sharing of information necessary for management decisions. In addition to the funds each country provides for research activities in support of management efforts, the PSC created two joint funds in 2001 to address additional information gathering needs and enhancement activities. The proposed projects are reviewed by independent scientists from the United States and Canada.

> "The future of cooperative fisheries management in the Columbia River basin will depend on the ability of the parties to formulate a common vision that addresses the interests of the parties and the resource."

While the CRFMP and the PSC shared the objective of rebuilding Chinook stocks by 1998, the salmon populations continued to decline in the early 1990s. Fisheries managers could control the harvest and hatchery production, but their influence on other activities affecting the survival of the populations and on the environment was minimal. There was no forum to adequately address the allocation of the conservation burden across the life cycle of the fish. The complexion of management changed when the Snake River sockeye salmon *O. nerka* were petitioned for protection under the ESA in 1991, and by 1994, a number of salmon populations throughout the Pacific Northwest were listed. The recovery plans required under ESA could provide an opportunity to comprehensively address the factors for decline, but to date, a plan acceptable to the region has not been produced. The large number of affected parties makes the task extremely difficult. The result is implementation of the ESA for salmon on sets of proposed actions rather than a cumulative effects approach.

Meanwhile the *United States v. Oregon* parties continue to negotiate a new management plan that addresses the conservation needs of the stocks while providing harvest opportunities. The parties entered into an interim 3-year management agreement while issues related to a long term arrangement are being discussed. One of the outstanding issues is the ability of the parties to commit to use their respective authorities to take actions that are consistent with the *United States v. Oregon* objectives in other processes. Stocks can not be rebuilt solely through actions in harvest and hatchery management.

In order to address concerns about weak stocks, changes in fishing patterns are already underway with the implementation of mark-selective fisheries. While mark-selective increase opportunities for increasing harvest of marked hatchery salmon, the tribes have raised concerns about the ability to accurately measure the effects on unmarked wild salmon with current methodologies. Mark-selective fisheries can discriminate at a gross level, but the ability to measure impacts on specific populations of interest may be compromised. New technologies are being considered to provide additional information for managers to consider. Passive induced transponder tags are beginning to provide insight on migration patterns in the river. Genetic sampling can provide additional information on stock composition of the harvest. Interagency cooperation and standardization will be necessary before new technologies can be applied on a broader scale.

“One of the most important characteristics of a succesful cooperative research project is the ability to formulate a common understanding of the question.”

The future of cooperative fisheries management in the Columbia River basin will depend on the ability of the parties to formulate a common vision that addresses the interests of the parties and the resource. The vision must be applied consistently across the activities of the life cycle of the fish for recovery and rebuilding to be achieved. In addition, the parties must make a commitment to the process and provide the necessary resources to implement the vision. The legal framework helps provide a structure for decision making.

Discussion Questions

What are the characteristics of successful and unsuccessful cooperative research projects? How do they differ on each coast, by scale?

One of the most important characteristics of a successful cooperative research project is the ability to formulate a common understanding of the question. From that common understanding, options with objective evaluation criteria can be generated. A catalyst is often needed to drive cooperation. In the case of the Columbia River fisheries management experience, litigation often served as the driving force leading to cooperation. The federal court maintains jurisdiction in *United States v. Oregon* and that accountability to a third party keeps the process going.

Due to the migration of the salmon, Columbia River fisheries management issues involve two countries, four states, tribal governments, and local government. Formulating a common vision is a challenge because of the different interests of the parties. The management process for Columbia River salmon consists of a series of linked processes that follow the life cycle of the fish. The key to making the linked processes work is to have continuity or overlap between the processes.

What kind of leadership is necessary?

Leadership needs to have a firm idea of their interests and how that fits into their vision of the future. Leadership also needs to be open and listen to the interests of the other parties. The mutual gains approach to negotiation opens up possibilities not considered when taking a positional or distributional approach to negotiation. Good communication skills are a must. In order to achieve a satisfactory resolution, leadership needs to address issues of substance, process, and relationship. Substance is important because it deals with the fairness of the resolution. Process is important because it ensures that all the relevant parties participate in a meaningful way. Relationship is important for the long-term stability of the resolution. Leadership needs to have the flexibility to consider actions as a learning experience and apply that knowledge in an adaptive management approach. Leadership needs to commit the necessary time and resources to develop the information to help resolve the management question. Leadership must be able to prioritize questions and information needs because of limited funding.

What challenges are commonly encountered? How have these obstacles been dealt with (e.g., regulatory requirements/permits)?

One of the challenges commonly encountered is a lack of a common understanding of the issue. This lack of a common vision can lead to confusion about what data needs to be collected to help address the issue. Another common challenge is a lack of commitment of resources, especially funding. In the current climate of budget reductions, basic monitoring and data collection often suffer. The length of the time series of data can lead to significantly different conclusions. The biggest substantive challenge is the recognition of the allocation of the conservation burden and how that it applied across the life cycle of the fish. Several attempts have been made to have that discussion at a regional level but have failed because of the strong positions taken by the parties.

How is local knowledge integrated into the scientific process? What roles are all parties playing throughout the scientific process?

Local knowledge is most often integrated into the scientific process through public hearings. In my experience as a biologist for the tribes, I have the opportunity to listen to tribal members convey the knowledge that has been passed down through the generations. The tribes take a holistic view of resource management, with an eye to providing for future generations. The challenge is bridging the traditional knowledge with scientific methods. Incorporating local knowledge in the decision making process validates the information and builds trust between the stakeholders, the scientists, and the policy makers. Building that relationship benefits everyone in the process.

“This lack of a common vision can lead to confusion about what data needs to be collected to help address the issue.”

The states, tribes, and the federal government are comanagers in Columbia River fisheries management. Each group maintains its sovereign management authorities and *United States v. Oregon* provides opportunities for the parties to discuss issues that are of mutual interest. Scientific issues are first discussed by representatives of the parties at TAC and PAC. The technical findings are then presented to the policy representatives. If a dispute arises at the technical or policy level, the issue may be brought to the court for resolution. The dispute resolution process was an important element in the CRFMP because it forced the parties to critically examine the pros and cons of a negotiated resolution versus a court ordered resolution.

What data quality activities work (e.g., peer-review, data management)?

The TAC process under *United States v. Oregon* is a forum where information relevant to in-river fisheries is reviewed by technical representatives of the parties. TAC develops a common data set for use by policy makers. A common data set has the advantage of providing the policy makers with a foundation to discuss options for management. The Chinook Technical Committee, the Selective Fishery Evaluation Committee, and the Data Sharing Committee provide the same function in the PSC process. The Chinook Technical Committee developed and maintains a model that is used to analyze past fisheries and the potential effects of proposed fisheries. It is important to standardize databases that have use over a broad geographic area. The cwt database is a good example of cooperative coast-wide data collection. The cwt database is maintained at central locations in the United States and Canada, which helps maintain data quality.

How are projects monitored for effectiveness?

The effects of management decisions are evaluated through a series of key indicators. Catch sampling provides information on the number of fish harvested and the contribution of specific stocks to the fishery. Escapement indicators measured at key locations provide information on how the number of spawners relative to the management objective. The information is then incorporated into analytical models that forecast the effects of management actions.

References

Lee, K. N. 1993. Compass and gyroscope: integrating science and politics for the environment. Island Press, Washington, D.C.

NPPC (Northwest Power Planning Council). 1986. Council staff compilation of information on salmon and steelhead losses in the Columbia River basin. Northwest Power Planning Council, Portland, Oregon.

ODFW and WDFW (Oregon Department of Fish and Wildlife and Washington Department of Fish and Wildlife). 2002. Status report: Columbia River fish runs and fisheries 1938-2000. Clackamas, Oregon.

Sohappy v. Smith, 302 F. Supp. 899 (D. Ore. 1969).

United States v. Oregon, 302 F. Supp 899 (D. Ore. 1969).

Yakima (Confederated Tribes and Bands of the Yakima Indian Nation) v. Baldrige, 605 F. Supp. 833 (W.D. Wash. 1985).

American Fisheries Society Symposium 52:141–143, 2006

Design, Execution, and Management of Cooperative Research

Tom Rudolph
Cape Cod Commercial Hook Fishermen's Association
210E Orleans Road, North Chatham, Massachusetts 02650, USA

At this point, few in fisheries would deny the importance and relevance of cooperative research. With a rapidly growing presence in the nation's fisheries, the time for that is past, and the challenges we now face are functional, not philosophical. Ironically, in many ways, the challenges find their roots in one of the greatest strengths of cooperative research: the partnership of industry, science, and management. These interests have had little experience working together, and while the money, the will, and the need are on hand, the tools and techniques are not. The rapid growth of cooperative research contributes to the challenge as well, especially when combined with the traditionally slow pace of management. In some cases, the pace of fishing for data has outstripped the capacity to process it and the maintenance capabilities of the research infrastructure. While stakeholders are well along in the process of working together in the field, in some ways, they lag in their ability to translate their results into management. Growing pains are to be expected though, and bright spots on the horizon are numerous. As stakeholders accumulate experience, both individually and as teams, the process will get smoother.

Successful cooperative research projects must feature strong partnerships—ideally, a team composed of industry, scientists, and managers is formed from the start and collaborates all the way from planning through execution to processing results. The Cape Cod Commercial Hook Fishermen's Association (CCCHFA) is especially committed to the concept of industry participation. Too often, the fisherman's role is limited to the charter of his or her vessel and these paper partnerships are not successful. They fail to achieve the full potential of cooperative research. While they may deliver a good product to the taxpayer or funder, and provide important supplemental income to the fisherman, they do not deliver the working relationships between stakeholders that form the third critical benefit of the process. CCCHFA involves fishermen from start to finish, taking direction from an internal industry-based research steering committee to plan and manage programs and working hard to get fishermen actually doing the science.

Flexibility and communication are extremely important in order to sustain and nurture these partnerships. Fishermen may have difficulty understanding the constraints imposed by the scientific needs of a project. Several times during the large, regional cod-tagging project, CCCHFA staff became pretty unpopular with our industry partners when we shut down our operations due to localized saturation. The fishermen could not understand why we did not want to tag while the tagging was good. However, once we demonstrated the problem through good communication, they demonstrated flexibility by shifting their thinking and moving forward in other directions. This is not a surprising problem—most fishermen have never dealt with a scientific protocol, statement of work, or research permit. Likewise, scientists, project managers, and funders may have never dealt with the limitations of vessels, the vagaries of weather, or the mysterious disappearances of fish and must maintain flexibility in their approaches to these challenges. Perhaps above all, the two sides must maintain communication and flexibility on matters of money. When budgets or business realities loom, each should try to stand for a moment in the other's financial shoes.

> "...ideally, a team composed of industry, scientists, and managers is formed from the start and collaborates all the way from planning through execution to processing results."

A final cornerstone of a successful project is accountability to the resource and, by extension, the public. Cooperative research projects should be held to a higher standard of biological sustainability and fiscal efficiency. CCCHFA promotes and practices an approach to project design that achieves these higher standards through three critical elements: full acceptance of project-related mortality; 100% retention and sale of legal, marketable catch with proceeds returned to project budgets; and payments to vessels using project-specific, fixed daily rates.

These objectives frequently present challenges, but all indicators point to their inevitability. For instance, in New England, many projects in 2004 were impacted by a National Marine Fisheries Service (NMFS) decision that severely restricted the availability of days at sea exemptions for research. Days at sea are the primary input control in New England, and many investigators were caught off-guard by this requirement to address project mortality, especially since it dramatically escalated vessel costs. Despite the challenges, CCCHFA believes that projects must be biologically sustainable through full application of input controls, hard total allowable catches, set-asides, or other means. Furthermore, they should not waste fish—target or bycatch species. While it may present obstacles, especially in the permitting process, legal-sized, marketable fish should be landed and sold. Finally, in order to avoid bias in data and perceived or actual impropriety in finances, vessels should return this fish revenue to project budgets and be paid fixed rates for their participation. If a vessel is allowed to keep fish revenues, the fishermen are more likely to alter their behavior to maximize catch and deviate from protocols. Fixed daily rates are essential to avoid controversies like those felt periodically in New England when fishermen, often but not always those outside the process, react with anger to tales of large grant payments and big fish checks, for the same trip, going in the pockets of other fishermen.

In New England, the main challenge faced by an investigator is usually permits—it has become increasingly difficult to secure permits on the same timeline as funding, and poor communication of shifting permit requirements often results in unnecessary delays, unpalatable changes to protocols, or outright rejections. CCCHFA typically submits permit applications before or along with funding proposals and still frequently has funding in hand before permits. Investigators face an uphill battle and must accept a much longer process if they want to land and sell the catch—in some ways, they are encouraged to leave it behind by discarding.

While timeliness (securing funding to answer pressing questions) and lack of abundance (failure to locate the fish one wishes to study) are additional problems often seen in New England, the one most frustrating besides permits is probably budgets. Once again, the main issue CCCHFA has confronted surrounds the landing and sale of fish. While our approach has been received positively in theory, implementation has been more difficult. In a nutshell, the federal government, the main funder of cooperative research, is much better at giving money away than taking it back. There is no workable method for NOAA Fisheries to collect and recompete or utilize these fish revenues. As a result, they have typically required investigators to estimate fish revenues and reduced the award or contract by this amount. This exposes contractors to significant risk if the revenue projections fall short and leaves them with three unpleasant choices—deficit spend the program, reduce expenses by scaling back on work, or alter the protocols in an attempt to increase harvest. Small entities like nonprofits or fishing businesses, or even state agencies and universities, need the security of a funding commitment that equals their budget if they are to deliver in full on their statement of work. Fish revenues have tremendous potential to save the taxpayer or funder money, fund additional research, or both, but they are not reliable enough to use as credit.

“In New England, the main challenge faced by an investigator is usually permits—it has become increasingly difficult to secure permits on the same timeline as funding.”

Solutions might best be found through the same cooperative approach that is yielding so much good science. For instance, it is clear that the permitting process is highly dependent on a healthy agency—staffing difficulties at the Sustainable Fisheries Division in New England have no doubt contributed to the delays in permitting. When CCCHFA led a congressional education visit to Washington, D.C. to tell the New England delegation how important cooperative research was, our delegation stressed the importance of maintaining funding for NMFS across the board. Cooperative research is highly dependent on infrastructure and support functions hosted by the regional offices, science centers, and councils. Moving our results into effective management would likewise be impossible without these partners.

The cooperative research functions hosted by the Northeast Regional Office, unique in the nation, also offer possible solutions to many of the challenges. The Cooperative Research Partners Program, because of its accessibility to managers, offers a rapid response to management needs that can yield very timely results. For instance, in response to a burgeoning haddock population and continued concern over cod, this funding arm was able to request proposals for projects that would test ways to maximize the haddock to cod ratio in experimental fisheries. Because these eventual fisheries would typically be granted exemptions from certain input controls in exchange for being held to a higher standard on catch, bycatch, and monitoring, these projects came to be known as Special Access Program Demonstration Projects. Real fisheries, sustainable and beneficial, have been created from these projects, and more are on the way.

“The process will work best if it is allowed to build upon itself—that is, if the working relationship and channels of communication are applied to each new challenge that arises.”

The process will work best if it is allowed to build upon itself—that is, if the working relationships and channels of communication are applied to each new challenge that arises. From data quality to management applications to pressing research needs, frequent consultations between industry, regional scientists both government and private, and managers will yield solutions in which all parties are vested. Regional Council Research Steering Committees are a good start but will require support and guidance, especially if burdened with heavy technical review responsibility. As always, time and money will impact this process as the workload increases. The revenues realized from fish caught in research can ameliorate this by providing additional funding to NMFS, provided a way can be found to keep it in house without exposing the contractors to risk. Priority should be given to this effort and all means utilized to explore it, including consultation with all stakeholders.

American Fisheries Society Symposium 52:145–147, 2006

Lessons Learned: Making Cooperative Research Work

John W. Watson
NOAA Fisheries Service, Southeast Fisheries Science Center
Post Office Box 1207, Pascagoula, Mississippi 39568-1207, USA

NOAA Fisheries Service researchers, commercial fishers, fishery organizations, and university Sea Grant organizations have undertaken several cooperative fisheries research projects in the past two decades to develop new fishing gear designs and techniques that reduce the impact of commercial fishing operations on endangered and threatened sea turtle species. These projects have been very successful in developing fishing gear and fishing practice modifications that significantly reduce the impact of commercial fishing operations on sea turtles while maintaining acceptable harvesting efficiency for directed species. While these projects have been successful, they have not been without problems, but the lessons learned from these projects have been very beneficial in designing and implementing more effective research partnerships in recent years.

I will discuss two specific projects in the context of the questions developed for the session *Lessons Learned: Making Cooperative Research Work*. The first project is the development of shrimp trawl design modifications to reduce the mortality of threatened and endangered sea turtles in penaeid shrimp trawl fisheries. The successful development and implementation of sea turtle mitigation technology in commercial penaeid shrimp fisheries around the world has been instrumental in the international effort to recover threatened and endangered sea turtle populations, while maintaining viable commercial shrimp fisheries. Research on shrimp trawl gear modifications to reduce the incidental take and mortality of sea turtles was initiated by the National Marine Fisheries Service in 1978 in response to mandates of the U.S. Endangered Species Act. Turtle excluder devices (TEDs), trawl modifications that allow the escape of sea turtles and other large animals from trawls, were developed and tested between 1978 and 1988, initially by a government research program and later under a cooperative program with industry. The government research program was successful in developing effective technology to reduce sea turtle mortality in shrimp trawls, and this technology was demonstrated to the industry in 1980. In 1981, voluntary use by fishers was encouraged through an intensive technology transfer project. The technology was not accepted by fishers due to the complexity and operational characteristics of the TED design. Research to refine and improve the operational characteristics was conducted between 1980 and 1984, but the technology was not voluntarily adopted by the industry. Although there were some fishers that provided design ideas and assisted in the government development program, for the most part, the industry was reluctant to work with government researchers, and there was general denial that there was a sea turtle problem in the fishery. Also, government researchers did not make significant efforts to involve the industry in the planning and conduct of the research. There was no effective communication between industry and government, and there was a general mistrust between the industry and government. In 1986, as the result of a mediation meeting between nongovernmental organizations and the industry, the government proposed mandatory use of the gear by the industry. The industry independently developed new designs between 1987 and 1989 that were evaluated for effectiveness in reducing turtle mortality and approved. Turtle excluder devices were implemented in

“While these projects have been successful, they have not been without problems, but the lessons learned from these projects have been very beneficial in designing and implementing more effective research partnerships in recent years.”

1989 through mandatory regulations but not without difficulty and adversarial confrontation between fishers and fishery managers. With widespread use, many operational and technical problems were identified. Between 1990 and 1992, fishers, industry organizations, and government and Sea Grant gear specialists initiated a cooperative research project that solved the operational and technical problems, and this effective cooperation continues today with new innovations being developed. Successful implementation required 5 years of intensive enforcement, refinement of technology, and technology transfer efforts. Through an effective technology transfer program, this technology has been successfully implemented in the U.S. domestic shrimp trawl fishery and 21 other countries worldwide.

The second project for this discussion is the cooperative government/industry project to develop sea turtle mitigation techniques for the commercial pelagic longline fishery. In 2000, estimates of turtle takes in the pelagic longline fishery raised concern that the pelagic longline fishery may be impacting the potential recovery of loggerhead and leatherback sea turtle populations in the Atlantic Ocean. In 2001, the NOAA Fisheries Service closed a significant portion of the North Atlantic to longline fishing by U.S. fishers and initiated a cooperative effort between fishers, universities, and the federal government. Components of the cooperative project included active participation of fishers and Blue Water Fishermen's Association in the design and implementation of the research, adherence to stringent scientific protocols, and significant input from fishers on experimental design components. In initial meetings between industry and other cooperative partners, there were problems relative to the need for stringent scientific protocols and the practical aspects of fishing practices and mistrust between the cooperative partners. A concerted effort was made to improve communications and trust and to reach acceptable compromise on disagreements. The result was that fishers were able to educate scientists with their knowledge of the fishery, turtle interactions, and potential practical solutions, and scientists were able to educate fishers on scientific research methodology and protocols. This cooperation, trust, and effective communication resulted in an extremely effective and successful research project. Between 2001 and 2003, the research program developed sea turtle mitigation techniques that demonstrated potential reductions in sea turtle interactions between 58% and 94% for loggerhead turtles and 44% and 86% for leatherback turtles, without significantly impacting target catch rates. The program also developed effective tools and techniques to safely remove fishing gear from sea turtles and other pelagic bycatch species, which has the potential to improve postrelease survival. The pelagic longline sea turtle mitigation technology has been implemented in the pelagic longline swordfish and tuna fishery in the Atlantic and Gulf of Mexico and in the Pacific swordfish fishery and is being evaluated in other pelagic longline fisheries around the world.

“Voluntary acceptance of new technologies requires financial or other incentives. Without incentives, acceptance of new technologies is not likely to succeed. New technologies that result in increased cost and/or loss of revenue will be resisted by users.”

These two examples of cooperative fishing gear research projects indicate some general characteristics of successful and unsuccessful cooperative projects. Although the first example was ultimately successful, it was a very difficult, lengthy, and expensive process and was not successful until a truly cooperative effort was attained. These experiences have resulted in the determination of several factors that are important to successful cooperative fishing gear research projects. First and foremost, the affected industry or constituents should be active participants in every aspect of research planning, technology development, and evaluation. Planning for new technology development and implementation needs to include a major long-term commitment for technology transfer and assistance. Voluntary acceptance of new technologies requires financial or other incentives. Without incentives, acceptance of new technologies is not likely to succeed. New technologies that result in increased cost and/or loss of revenue will be resisted by users. Mandatory requirements for new technologies must include effective enforcement, and regulations should be as flexible and easily modified as possible to allow successful implementation and improvement of new technologies. Most importantly, successful cooperative research programs require effective communication, cooperation, and trust between cooperative research partners.

Effective leadership is also a vital component of successful cooperative research programs. The most effective programs include leadership that is effective in communicating with research partners and the development of advisory groups with the right mix of expertise and skill to be effective. Leadership must be receptive and proficient in developing compromise solutions when partners have divergent points of view or be firm and resolute when compromise is not an option. A representative industry organization that can effectively communicate industry concerns, needs, and opinions in negotiations and that can form an effective technical advisory group is especially beneficial in a successful cooperative program.

“Partners must be effective communicators and negotiators willing to compromise to move the process forward.”

Challenges that are commonly encountered include lack of trust and understanding between program partners and misinformation. Other significant problems in developing cooperative research projects are securing adequate funding, Endangered Species Act and exempted fishing permits requirements, including National Environmental Policy Act and environmental impact statements, and other regulatory and permit requirements that can significantly delay and, in some cases, prevent research programs.

Cooperative research programs have proven to be highly effective in developing solutions to critical fisheries problems. To be successful, they must have common goals and sincere commitment from all partners. Partners must be effective communicators and negotiators willing to compromise to move the process forward. Programs need effective and expert advisory personnel to assist with regulatory and permit requirements, ensure that local knowledge is integrated into the scientific process, maintain acceptable scientific standards, and effectively develop and execute a successful research program.

American Fisheries Society Symposium 52:149–153, 2006

Session 2B: Lessons Learned

Making Cooperative Management Work Panel Summary

Paula Cullenberg
Alaska Sea Grant Marine Advisory Program
1007 West 3rd Avenue #100, Anchorage, Alaska 99501, USA

Cooperative management—regardless of what it is called—you know you have it if there is a commitment to work together, share information and communicate to reach decisions through consensus. [Bill Mattes, Great Lakes Indian Fish and Wildlife Commission]

The second session of the joint Sea Grant/American Fisheries Society symposium focused on the lessons learned from cooperative fisheries research and management. Paula Cullenberg with the Alaska Sea Grant Program moderated the panel. Below are some speaker highlights, followed by a summary of the facilitated panel-audience discussion. White papers prepared by the panelists follow this panel summary.

Ted Hoskins, minister with Maine Seacoast Mission, memomber of the Saltwater Network, Penobscot East Resource Center, Stonington Fisheries Alliance, Downeast Groundfish Initiative

- There are two resources of concern—marine resources and human resources and how they interface. We need to be equally interested in how we manage the human resource. These communities need maintenance within the larger fisheries context.
- How do we get people to become stewards of the resources? Ensuring compliance is impossible without stewardship. There are not enough marine patrols in the world to follow and chase people to make sure they follow the rules.
- How do we ensure sustainability—in fisheries *and* in fishermen? This should be a priority for community-based actitivities.
- In the Downeast Groundfish Initiative, it was discovered that a set of principles to guide them was necessary from the beginning, basic principles that include protecting habitat, spawning grounds, juveniles, food chains, recognition that fish are a public resource, and protecting the community and the rights of small boats to have access to this resource. A lot of attention needs to be given to developing guiding principles from the onset.

- Capacity building through education and making materials available to local groups in forms they can use promotes access to cooperative research. This provides avenues for local involvement and teaches the skills that they need to participate in fisheries management.

Summary and Overview of Panel-Audience Discussion

At the completion of the presentations, the moderator facilitated a discussion between panelists and audience members to address the lessons learned from cooperative fisheries management, including characteristics of successful projects, challenges, impacts, and the future of cooperative fisheries management.

Characteristics of Successful Cooperative Management

Cooperative management is generally characterized by consensus, accountability, communication, and cooperation. As such, it can often be difficult to achieve but carries benefits such as developing stewardship among users of a resource and sharing local knowledge that makes science more effective. Cooperative management has the potential to reduce the need for enforcement actions, which are often limited by financial resources. While many fisheries management regimes are top-down in their approach, cooperative management regimes are either bottom-up or somewhere in between. Several examples of different organizational regimes were presented.

Many cooperative management processes, although not all, are legally mandated, such as the

Tom Iverson, Columbia Basin Fish and Wildlife Authority

- The Columbia Basin Fish and Wildlife Authority (CBFWA) was created to coordinate efforts to protect and enhance fish and wildlife resources in the basin and to help in implementing the Northwest Power and Conservation Council's Fish and Wildlife Program.
- The CBFWA coordinates 19 fish and wildlife agencies and tribes and is a consensus-based organization with different levels (local and regional) engaged to provide policy guidance.
- Most of his time is spent managing people, facilitating comanagement decision making, formulating policy issues, and providing support for conversations around important topics, which do not always result in decisions.
- This region is having difficulty balancing bottom-up, locally led planning, with top-down regional planning.

- Written arrangements were set up to coordinate management within the Great Lakes region. Each arrangement is different, but in all arrangements, they recognize that their tribal rights to use the resources come with the responsibility to manage resources.
- "They operate based on consensus, accountability, and cooperation. Cooperative management is sometimes referred to as coordinated, joint or shared—but regardless of what it is called, you know you have it if there is commitment to work together, share information, and communicate to reach decisions through consensus. You have to work together if you are working with a shared resource."
- The founding document is a strategic fisheries management plan that is a non-binding agreement. Consensus, accountability, communication, and cooperation are outlined in the treaty.

Bill Mattes, Great Lakes Indian Fish and Wildlife Commission

Great Lakes Indian Fish and Wildlife Commission, which developed from treaties between the Ojibwa tribes and the U.S. government around resource management in Lake Superior. Treaty rights were reaffirmed by the courts in 1971, and since then, written arrangements have been drawn up to coordinate management. As Bill Mattes noted, "the tribes recognized that a braided rope is

Roy Stein, professor, ecology and evolution, The Ohio State University

- Partnership as a continuum—small scale to large scale.
- Why do partnerships work? Flexible administration on both sides helps. All partners involved want to answer the research question.
- The bottom line: the people are most important; the ones that interact are the ones that make it work.

stronger than individual strands." By forming into the commission and hiring their own scientists, the tribes developed a means to keep the traditional thought and way of managing resources integrated with the western science approach to management. The backbone of the commission is the belief that protecting the resources for future generations is of paramount importance.

The Columbia Basin Fish and Wildlife Authority presents another level of cooperative management, as addressed by Tom Iverson. This group coordinates all of the regional fish and wildlife managers to protect and enhance the fish and wildlife resources in the entire basin. The CBFWA generates consensus recommendations for mitigating the impacts of the Federal Columbia River power system.

In Maine, fishermen have developed a three-tiered organizational process, which provides networking and representation at each level leading to an increased voice in fisheries management. The Saltwater Network (a cross-border organization) tries to support middle-tier groups, called resource centers (such as the Cobscook Bay Resource Center), which in turn, support more local groups of fishermen.

As Ted Hoskins noted, groups attempting cooperative management first need to establish a set of principles to guide them; without this, the group will be representing only their interests. Basic principles might include goals such as protecting habitat, enhancing spawning grounds, or protecting the local community's access to the resource. Bill Mattes also pointed out that the Great Lakes Fishery Commission, which acts as a comanagement body among all governments with fisheries management authority, uses a strategic fisheries management plan as a founding document. These examples suggest that the organizing documents used by the cooperative partners are very important in solidifying their goals and keeping the broader overarching mission at the forefront of their projects.

As Paula Cullenberg mentioned during the panel's discussion, having a community develop their plan and approach from the ground up will often enable regulators to respond more positively. On the Columbia River, local stakeholders are involved in the Endangered Species Act recovery planning. This allows the stakeholders that are regulated by the eventual plans, to participate in their construction.

Although it is clear that there is not a one-size-fits-all formula for cooperative management, a number of common practices enable this form of resource management to be worth the hard work and time needed to develop and maintain such a system. On the other hand, it is important to leave room for creativity in the expectations and goals. Set expectations carefully, try to be flexible while keeping the end goal in mind. If you do not preset your expectations on how to manage and coordinate in a cooperative manner, you may end up with better solutions and outcomes that you had never anticipated. Be patient, as building relationships and cooperation take hard work and commitment. If you think you are going to come to resolutions everyday, you will be disappointed. Nonetheless, all of the steps along the way are constructive. Twenty years ago, cooperative fisheries research and management issues would not be discussed at the American Fisheries Society annual meeting. The fact that this symposium occurred is a testament to the progress that has been made.

Challenges of Cooperative Management

Several challenges were mentioned by the panelists, including time, effort, funding, maintaining motivation, communication, and engagement. Challenges to successful cooperative management are best summarized by a sentiment shared by the panelists—"it takes a lot of work." Funding was discussed as one challenge. Roy Stein noted that his experience with successful projects was due in part to assured funding from the state of Ohio. Ted Hoskins also said that in Maine, groups working to include fishermen must also raise their own funds to stay involved. Finding adequate funding is always a challenge.

Keeping people, members, or groups involved and the motivation high is challenging. As Ted Hoskins noted, at the very local level, many fishermen do not want to participate in group meetings; "they want to be independent and do not want to go to any meetings." Tom Iverson noted that much of his work in the comanagement arena involves facilitation, managing people, and trying to generate conversations, "some of which do not always result in decisions." It is difficult for local stakeholders to make the connection to regional planning exercises and for regional planners to balance the needs between local stakeholders—prioritizing one watershed over another.

Building the collaborative skills and resources of community members is an important part of successful cooperative management. In Maine, capacity building starts at the local level so that fishermen can learn the skills they need to participate in the fisheries management process. It is important for everyone to have the scientific tools and build a skill set necessary so that they can feel equal and effective in the management arena.

Oftentimes, programs start small and then grow. Other examples suggest that certain programs work most effectively at smaller scales. The CBFWA started by recommending specific projects in the 1980s that ultimately led to the development of subbasin level conservation plans. As Roy Stein noted from his experience in Ohio, cooperative research can also lead to better cooperative management.

Impacts

Cooperative management can play an important part in supporting fishing communities, particularly in rural sites where alternative income sources can be limited. By recognizing the human dimension of resource management, enforcement efforts can be reduced and stewardship increased. For example, scientists were hired to argue that tribal cases needed to include the values of the Indians in their analyses and decisions. By listening to community members and elders, to the stories passed down and their reliance on and connectiveness to the resource, scientists can better plan their work. By listening to what commercial fishermen are seeing and what they think is going on, scientists can use that knowledge in their research design. Listening is critical to true integration of local knowledge.

What's Next?

While cooperative management is growing in use, it was noted that agencies, such as NOAA Fisheries, the U.S. Fish and Wildlife Service, or state or tribal management organizations generally have the ultimate authority. Stakeholders providing recommendations on an advisory level is currently the norm. Greater devolution of authority and broader community management authority may develop in the years to come with continued capacity building and the recognition of the merits of the cooperative approach. As Bill Mattes noted, in tribal arrangements, the courts often play a significant role in defining authority. However, if you have good cooperation, you may have less court-directed management.

American Fisheries Society Symposium 52:155–158, 2006

Some Opinions Regarding What Makes Collaborative Management Work

TED G. HOSKINS
Post Office Box 931, Blue Hill, Maine 04614, USA

What Are the Characteristics of Successful and Unsuccessful Cooperative Management Projects? How Do They Differ on Each Coast, by Scale?

It is tempting to follow along with the spate of business management books that list "Ten Habits of Successful..." and inhabit the bookshelves of many a would-be manager. This would not be all bad, for there are, without question, characteristics that surface in almost every successful project, and many certainly find their rightful place in fishery management. As to the determination of "successful or unsuccessful," that is an uncertain task in itself, often depending on whom and when you ask the question. There are unintended results that at times turn out to be the real winners!

Here are a *few* characteristics of what I see as successful projects:

- *A well-articulated and broadly accepted goal* is essential. This requires a tremendous amount of on-the-ground work so that all recognize in the goal some of what they want. Herein lies the basis for inclusivity both in defining the goal and in broadcasting its purposes.
- *Access to process and decision making* by all affected parties. This does not mean that everybody has to serve on a committee or be directly represented. It does mean that all need a reasonable point of entry into the process that leads to decisions. (For lobster management in Maine: the Department of Marine Resources, the Lobster Advisory Council, and lobster zone councils [A–G] and associated zone districts[1])
- *Balance in decision-making power* goes beyond access and means that everyone has an honest and equal opportunity to affect decisions. Without this, there is disillusion and disaffection. This is one of the more difficult areas in natural resource projects that deal with multiple jurisdictions, not to mention a skeptical public.
- *Clearly defined boundaries*—geographic, political, economic and social—are important for effective dialog. In some ways, this is a refining of the goal in specific arenas.
- *Funding.* Lack of funding is noted as a major factor in projects that do not succeed. So careful planning is essential, not just to get started, but to keep going. While the number of projects keeps growing, it is increasingly obvious that the number of funders does not keep pace. Col laboration and networking are key to long-term success, along with a willingness to adapt to changes in funding patterns such as the emphasis on regional planning.

Let us just list some more characteristics:

- Inclusive ... starting with the harvesters
- Adaptive...in a wide variety of ways
- Persevering... and then some!
- Politically savvy ...maybe even connected!
- Communicative... in every direction
- Strength of leadership... essential
- Strength and breadth of board of directors to meet varying needs

[1] There are seven lobster zones (Zones A–G) along the coast of Maine, and each has 7–13 subunits, called districts. Districts represent smaller geographic areas, typically particular harbors and ports. They were established by state legislation and have been delegated special authority over lobstering management activities within their areas. This is an example of area-based comanagement.

- Effective use of media
- Open to collaboration and networking at varying levels

What Kind of Leadership Is Necessary?

It is difficult to say too much here without knowing the nature and size of the project. Many of the same characteristics just noted in a successful project are also to be found in successful leadership: good communication skills, a willingness to collaborate, more than a little political savvy, a willingness to listen to new ideas and different perspectives, a reputation for being approachable, and being filled to overflowing with optimism and perseverance.

Leadership almost always centers in a single individual but, in the same measure, works best when explicitly shared with a board or committee. The executive director, by whatever name he/she is given, needs the drive and perseverance, the energy, and the skills to keep the project on the move and on track. One of the more important skills is knowing how to let go some of the work that must be done…put the board to good use according to their abilities and interests. People do not serve on a board because they have extra time… they have talents that need to be used or they will feel useless! A board is primarily essential for breadth of perspective in sharing decision making and the praise/blame that goes with those decisions, not to mention saving the executive director from burnout. A well-balanced board that is put together with careful planning is an invaluable asset for community support. Board training and education programs, yes even trips to visit other projects, can bring surprising new life to an organization. Leadership should not be a life sentence for the director or the board. Develop and systematize ways to rotate the positions of leadership; this is important for capacity building as well as maintaining interest. In Maine, most projects are identified with a well-known, strong leader, and one wonders if they could survive a change in leadership. The "founder-syndrome" can be a real problem as a project matures and adapts.

> "The executive director...needs the drive and perseverance, the energy, and the skills to keep the project on the move and on track."

What Challenges Are Commonly Encountered, and How Have These Obstacles Been Dealt With?

We all know that funding is a major challenge. How much of an organization's time and energy are consumed in securing funding!? We, in Maine, have been very slow in finding ways to shift a portion of the financial burden to the users who benefit, not only fishermen, but also communities and institutions. Yet this shift is necessary. We have begun to talk at town meetings and local nonfishing organizations about supportive long-term funding of the projects that benefit the community as a whole. This has met limited success.

The Downeast Groundfish Initiative has recently convened, with the help of Saltwater Network, a group of leaders of nonprofits that have traditionally been competitors for the same funds. Following on the study done for the Barr Foundation, "Network Power for Philanthropy and Nonprofits," a collaborative group, is slowly finding its way toward developing a new level of trust among the participants, who have all known and worked with each other for years, and then intentionally looking at what needs to be done and talking seriously about who can best do it and supporting that proposal, rather than all going after the same piece of the pie. It is too early to tell how this will work, but it is, for me, a hopeful venture.

All our talk of community-based this or that is in danger of being fraudulent if we do not find ways to effectively involve fishermen and local citizens. Too often, it is nonindustry middle-management, organization-type people who are doing everything. The big challenge is getting down to where the people are. Saltwater Network, Penobscot East Resource Center, Cobscook Bay Resource Center, and Northwest Atlantic Marine Alliance are all committed to involving the people at the local level, such as in the Stonington Fisheries Alliance, but it is no easy task. As noted elsewhere, developing local alliances, industry groups, and fishermen's organizations is a slow, demanding, yet essential task, with success being uncertain and elusive.

How Are Local Knowledge and Interests Integrated with National, Regional, and State Interests?

This may well be the most difficult and yet most promising of all the questions. I can only speak to what is happening in the East, Maine in particular. The National Marine Fisheries Service management system is organized to work in classic top-down style. Though there is lip service given to industry involvement, only larger boats and corporately owned vessels have the financial ability to effectively participate in the complicated and seemingly endless processes that lead to decisions. Smaller owner-operators feel completely disenfranchised.

In Maine, and the wider Gulf of Maine, we have been developing what we call resource centers (RCs), which are nonadvocacy organizations that have a special role to play between state and regional councils on the one hand and local boat owners and communities on the other. A primary role is to encourage and support the development of local fishermen's organizations, not an easy task in independent-minded coastal Maine. This requires a tremendous amount of capacity building over a lengthy period of time. It also requires the filtering and funneling of pertinent and useful information from regulatory bodies to the local groups in a form that invites reaction and response. The RCs "resource" the local groups in a multiplicity of ways: geographical information system services, mediation services, public education services, collaborative science connections, historical perspectives, funding support, technical support, and so forth. The RCs play a significant role in getting the local knowledge and interests of fishermen into the discussions and decisions of the regulatory and management bodies, without themselves being the messengers.

Within the lobster industry in Maine, a form of comanagement has made all the difference by dividing the state waters into seven zones, with districts at the wharf level, then representative councils in each zone, and last a lobster advisory council that meets with the Department of Marine Resources and advises both the commissioner and the legislature. There are a limited number of issues (number of traps on a trawl, hours of fishing, number of traps fished, etc.) that they can address. After a number of years of slow and uneven capacity building, this zone system, which recognizes the varying fishing needs and patterns on a long coastline, has been quite successful.

> "We all live under the constraints and demands of goals and objectives, measurables, and outcomes. Our funding stream is usually dependent on successfully evaluating the project in terms of its predetermined goals."

How Are Projects Monitored for Effectiveness?

We all live under the constraints and demands of goals and objectives, measureables, and outcomes. Our funding stream is usually dependent on successfully evaluating the project in terms of its predetermined goals. Most projects are monitored in these time-honored ways. However, there are other ideas afoot. In Saltwater Network (a cross-border group working for the ecological health of the Gulf of Maine), there is experimentation with a form of evaluation that is ongoing throughout the project, not just at its conclusion. It is aimed at reflective assessment of what is happening at regular points throughout the life of the project so that everyone (both funder and project) learns as they go, lessons learned can become the basis for new activity during the project, goals may be redefined, and all learnings are shared broadly with others. This time-sensitive, adaptive evaluation and monitoring promises a much shorter feedback into the loop of management decision making.

American Fisheries Society Symposium 52:159–162, 2006

Making Cooperative Management Work: Columbia Basin Fish and Wildlife Authority

Tom Iverson
Columbia Basin Fish and Wildlife Authority, Pacific First Building
851 SW 6th Avenue, Suite 260, Portland, Oregon 97204, USA

The Columbia River flows for more than 1,900 km and drains 670,000 km^2 in seven states and Canada and covers an area approximately the size of France. The Federal Columbia River Power System (FCRPS) consists of 29 hydroelectric and irrigation dams. In addition to the FCRPS, there are more than 1,200 additional dams on the Columbia River and its tributaries for the purpose of hydroelectric generation and irrigation. The construction and operation of the FCRPS has directly impacted fish and wildlife resources in the basin, with several fish species extirpated and other fish species listed as threatened or endangered. The secondary effects of the FCRPS due to irrigated lands, development, and population increases have had significant impacts on wildlife, anadromous and resident fish species.

In 1980, the U.S. Congress passed the Pacific Northwest Electric Power Planning and Conservation Act (NPA) "to encourage, through the unique opportunity provided by the FCRPS … to protect, mitigate and enhance the fish and wildlife, including related spawning grounds and habitat, of the Columbia River and its tributaries, particularly anadromous fish which are of significant importance to the social and economic well-being of the Pacific Northwest and the Nation and which are dependent on suitable environmental conditions substantially obtainable from the management and operation of the FCRPS and other power generating facilities on the Columbia River and its tributaries." The NPA created the Pacific Northwest Electric Power and Conservation Planning Council (recently renamed the Pacific Northwest Power and Conservation Council), a four-state compact, with appointees by the four governors of Washington, Oregon, Idaho, and Montana, directed to (1) develop a regional conservation and electric power plan, and (2) develop a program to protect, mitigate, and enhance fish and wildlife and to otherwise expeditiously and effectively carry out the council's responsibilities. The NPA calls for the council to rely on recommendations from the fish and wildlife managers to develop the program.

> "When the role of the cooperating parties is clearly understood and timelines and outcomes have been clearly identified and agreed upon, progress can be made."

The Columbia River Basin Fish and Wildlife Program is funded by Bonneville Power Administration (BPA) through wholesale electric rates charged to customers of the FCRPS. The program is currently funded at approximately $145 million annually to support a broad array of projects intended to protect, mitigate, and enhance fish and wildlife populations. The program funds more than 450 projects that support hatcheries, habitat acquisition and restoration, passage improvements, research, monitoring and evaluation, coordination, and other programs. The council makes project recommendations to BPA, and BPA funds the projects consistent with the council's program.

The state and federal agencies and Indian tribes of the Columbia River basin are the legally recognized managers of the fish and wildlife resources, through federal and state statutes, treaties, and court actions. In 1987, these fish and wildlife management agencies and Indian tribes joined together to form the Columbia Basin Fish and Wildlife Authority (CBFWA) for the purpose of coordinating their efforts to protect and enhance the fish and wildlife resources of the Columbia River basin in the implementation of the Northwest Power and Conservation Council's Fish and Wildlife Program. The three organizational objectives as outlined in the CBFWA charter are to provide interagency fish and wildlife management coordination, provide a centralized, regional entity for Columbia Basin Fish and Wildlife Program coordination, and coordinate activities between CBFWA and the land and water management authorities of the Columbia River basin (CRB).

The CBFWA is a consensus-based organization whose membership is composed of the state fish and wildlife management agencies of Oregon, Washington, Idaho, and Montana; the two federal agencies, U.S. Fish and Wildlife Service and NOAA Fisheries; and the 13 Indian tribes in the CRB. The CBFWA "Members" are the state agency directors, federal agency regional directors, and chairs of tribal governing bodies or duly authorized representatives of such directors or chairs. The Members establish policy direction. The Members meet several times annually to discuss and reach consensus on issues of common concern facing the fish and wildlife managers in the CRB. The Members direct the activities of the executive director, staff, and its committees. The Members Advisory Group consists of senior managers active in CRB management. They assure that the policies and directives of the Members are carried out and make recommendations for policy consideration. The CBFWA has technical committees to provide technical recommendations or analysis to the Members or the Members Advisory Group. They receive their assignments from the Members or the Members Advisory Group.

A role of CBFWA has been to provide policy and technical support to the council and BPA in their decision making. The role has varied depending on the clarity in vision of the program. In the early stages of the program, the CBFWA members provided guidance through measures in terms of projects (actions on the ground). As the program matured, the measures became less project specific. In the early 1990s, CBFWA coordinated activities for agencies and tribes to prepare an integrated system plan (coordinated subbasin level plans) and a comprehensive environmental analysis (essentially a programmatic National Environmental Policy Act). These plans provided context for implementation of projects through the council's program. It has also been the role of the CBFWA to select and prioritize projects against the measures identified in the program. With the development of community-based planning in the basin, the role of a regional fish and wildlife entity in planning and project selection is diminishing. The regional framework for the subbasin planning efforts is not clearly articulated, and therefore, CBFWA's current role is not well defined.

As an example, in 1991, CBFWA developed subbasin plans for all the anadromous fish watersheds in the CRB. The plans were successfully completed, and many of the actions recommended by the fish and wildlife managers were implemented. However, the plans were not perceived by policy makers as adequate due to the lack of involvement of local stakeholders, since not all local interests had participated in the planning process. In a workshop in 2000, the CBFWA members acknowledged the importance of including local land owners in the planning process since, in most cases, the areas of greatest need for habitat restoration are on private property and out of the fish and wildlife manager's direct influence. With the prolific creation of watershed councils in the CRB, the role of a regional entity, in this case CBFWA, guiding project selection at the local level came into question. The local watershed groups desired a stronger role in selecting that projects to implement in their watersheds to address locally prioritized limiting factors.

> "In the absence of a clear vision, cooperative groups will engage in displacement behavior and fail to accomplish useful deliverables."

The response to the concerns raised by the local groups was for the council to lead an effort to develop subbasin plans by local stakeholders. The CBFWA as an entity was discouraged from participation in this planning effort, although the individual CBFWA members participated in the development of these plans at the local level. The focus of this subbasin planning exercise was to build from the bottom up; however, the mandate of the NPA to be consistent with the plans of the fish and wildlife managers appears to have been lost in this exercise. The role of the regional coordinating body has not been defined by the regional decision makers.

In the CRB, recovery planning under the Endangered Species Act and subbasin planning under the NPA are relying on local involvement and decision making. This is consistent with the new paradigm in natural resource management where we allow the regulated to assist in developing the operational and regulatory guidelines. However, in this large regional program, the balance between a bottom-up, locally led planning effort and a top-down regional plan is not being well orchestrated. It is very important for comanagement to be successful in the CRB, that a clear regional framework be established that guides the implementation of the locally created subbasin plans. The tension between regional priorities and local priorities must be balanced in a open and transparent process.

> "Leadership must be committed to the comanagement process, even if they are not the individual representatives from their agency involved in the comanagement process."

The CBFWA provides a useful model for evaluating cooperative management. The organization and its members have been through several changing political and social environments. It is apparent that when CBFWA has been most successful, the vision, objectives, and timelines have been clear and well defined. The CBFWA has struggled when that clarity has been compromised.

Discussion Questions

What are the characteristics of successful and unsuccessful cooperative management processes?

A common vision and purpose for cooperative management provides successful outcomes. When the role of the cooperating parties is clearly understood and timelines and outcomes have been clearly identified and agreed upon, progress can be made. In the absence of a clear vision, cooperative groups will engage in displacement behavior and fail to accomplish useful deliverables.

Within CBFWA, our members have been able to develop an integrated system plan that coordinated 37 subbasin plans in 1991. The plans identified objectives and actions for each anadromous subbasin. The planning exercise forced the members to agree on habitat and hatchery strategies within each of the subbasins for a comprehensive regional plan. Although this planning effort was successful, history has shown that, over time, the political landscape had changed enough that repetition of the planning effort was required only 10 years later.

When participants in cooperative management have conflicting interests, with no room for compromise, success is rare. Within CBFWA, our members deal with water allocation issues on a regular basis. Water is needed in the upper basin to support resident fish species via stable reservoir levels, while water is needed downstream to support anadromous fish migration via spill and flow. In the past, the CBFWA membership has been able to resolve these differences and provide a compromise operations plan via the development of the Watershed Equity Team Report in the late 1990s. This plan was scrapped due to a federal agency's obligation for ESA protections on anadromous fish and failure to find compromise for resident fish needs under the new regulatory environment.

What kind of leadership is necessary?

Leadership must be committed to the comanagement process, even if they are not the individual representatives from their agency involved in the comanagement arena. In CBFWA's early years, our member agencies had ownership of the institution and its products. Recently, the members have dealt directly with regional decision makers as individuals and have viewed the CBFWA role in a lesser light. Technical level participants have not communicated well with their superiors (or vice versa), and confusion has occurred on what products and methods CBFWA members were engaging to address regional issues. The reasons for this change in commitment from our members are varied and legitimate, but the point is that the agencies and tribes reduced their commitment to the comanagement process without explicitly acknowledging so. This reduced commitment is expressed by sending technical level staff to participate in policy level meetings.

It takes many small decisions to develop resolutions to complex problems. An effective leader will make the small decisions and move larger concepts forward. Many efforts are stalled out in deciding the shape of the table, rather than on the potential outcome of the collaborative effort. While some of the smaller decisions are important for the group to address, a strong leader will not allow the group to linger too long on less important issues.

What challenges are commonly encountered, and how have these obstacles been dealt with?

There are many challenges that obstruct productive comanagement (i.e., self interest outweighing mutual interests, conflicting interests that cannot be resolved, and unrealistic expectations), which is precisely why there is a need for coman-agement. Issues outside of CBFWA's sphere of influence have drawn participation of our members away from the CBFWA table. For example, a major challenge for CBFWA revolves around water allocation between upriver flow control for resident fish species and downriver flow and spill for anadromous fish migration, as described above. Although, within CBFWA, we have been able to reach solutions on this issue, the intense polarized regional politics around this issue have prevented a solution from being implemented.

"Many efforts are stalled out in deciding the shape of the table, rather than on the potential outcome of the collaborative effort. While some of the smaller decisions are important for the group to address, a strong leader will not allow the group to linger too long on less important issues."

For anadromous fish, the hatchery versus wild debate has had the potential to stifle actions within CBFWA. The solution to this dilemma has been to resolve issues one decision at a time, on a case-by-case basis. Hatcheries are permitted in some subbasins and not in others. Production hatcheries are permitted in some subbasins, while supplementation facilities are exclusively tested in other subbasins. This solution allows a flexible approach and ample room for compromise.

How are local knowledge and interests integrated with regional, state, and tribal interests?

The CBFWA relies heavily on vertical integration/communication among our members. We operate at both a technical and policy level, and we expect our members to communicate within their agencies and tribes to best engage in finding resolution on issues. When the vertical communication is interrupted, we have difficulty addressing issues (i.e., misunderstanding at the policy level of what the technical workgroups are accomplishing and misunderstanding at the technical level of what the policy folks require for decision making).

In most cases, the local biologists from the state, federal, and tribal agencies are engaged in community watershed planning groups. Those biologists bring the local perspective to the regional dialog through their interaction with their supervisors. It is very important, and sometimes difficult, to balance the top-down guidance with locally led initiatives. This is the current challenge in the CRB and will become the focus of the program for the next several years.

American Fisheries Society Symposium 52:163–168, 2006

A Discussion of Cooperative Management Arrangements within the Ojibwa Ceded Territories

William P. Mattes and Neil Kmiecik
Great Lakes Indian Fish and Wildlife Commission
100 Maple Street, Post Office Box 9
Odanah, Wisconsin 54861, USA

In the Ojibwa[1] ceded territories, cooperative management arrangements are normally founded in court decisions reaffirming treaty-reserved rights, with each arrangement differing in various ways. In all cases, the Ojibwa, as governments and as nations, view that harvest rights carry with them the responsibility to protect and manage resources.

Background

From 1836 to 1854, four land cession treaties were signed between the United States Government and the Ojibwa tribes living in a vast area around Lake Superior that later became the states of Michigan, Wisconsin, and Minnesota. In return for their land, the Ojibwa reserved the right to hunt, fish, and gather throughout the ceded territory, including Lake Superior. As long as resources were plentiful, this reservation of rights seems to have been of little concern. But as fish and wildlife populations began to decline due to increasing exploitation and habitat destruction, the states began to develop regu latory programs and to enforce their laws against the Ojibwa. Nevertheless, survival and tradition led tribal members to continue to hunt, gather, and fish either on reservation lands or off the reservation where they were subject to arrest by state wardens. And so only a few decades after the last treaty was signed, the Ojibwa began to be denied their treaty-reserved rights.

Starting in the 1970s and continuing through today, tribal members and tribal governments began challenging the authority of the states to apply their resource regulations against tribal members hunting on ceded lands and fishing in ceded waters, both inland and Lake Superior. In a series of federal and state court decisions, the treaty-reserved rights of the Ojibwa were reaffirmed (e.g., in Michigan, the 1971 Jondreau Decision; in Wisconsin, the 1972 Gurnoe and the 1983 Voigt decisions; and recently during the 1990s in Minnesota, the Mille Lacs and Fond du Lac decisions). In some instances, after the existence of the rights was reaffirmed, further litigation followed to establish the scope of state regulation, the adequacy of tribal regulations, and how the resources should be allocated. In other cases, further litigation did not follow and the parties chose to negotiate regulations and allocation issues on a periodic basis. Also, in some cases, management authority was an issue. Different types of written agreements, approved by the courts, were used to set up institutional arrangements to coordinate management.

Development of Tribal Management Capabilities and Infrastructure

Soon after each court decision was made reaffirming the rights, a basic question that tribes pondered was how should the rights be implemented. The tribes signatory to the 1842 treaty looked to the northwest and saw that those tribes had formed an intertribal commission. The Ojibwa recognized that a braided rope is stronger than the individual strands of fiber and formed their own commission, called the Great Lakes Indian Fish and Wildlife Commission (GLIFWC).

[1] Anishinabe (although spelled and pronounced several ways) is the correct name for the people since this is the term they use to describe themselves. Ojibwa is what other tribes called the Anishinabe; and Chippewa is what non-Indians called the Ojibwa (Benton-Banai 1979:94–102).

The Ojibwa leaders also recognized that they needed their own scientists because those were the people that non-Indians and the courts turned to for answers to biological questions. However, in forming their commission, the leaders stated that traditional Anishinabe culture and values were to be infused in all aspects of its work. In the 1992 GLIFWC Strategic Plan, they explained that "The 'Anishinabe Way' underlies the unique approach to resource management which is brought by tribal people into the critical, modern day decisions, regarding natural resources. Traditional thought directs management to be holistic and integrated, respectful of all creation. An understanding of the universal order and recognition of man's dependence on all other life forms, rather than his dominance, assures holistic management. Traditional thought also demands long-term vision, protecting the well-being, not just of the next generation or two, but of the 'Seventh Generation,' thus extending responsibility for the impact of management decisions far into the future" (Great Lakes Indian Fish and Wildlife Commission 1992).

"...the Ojibwa hired their own fisheries and wildlife biologists, in part to review data gathered and analysis done by state biologists and to compile accurate resource inventories."

So the Ojibwa hired their own fisheries and wildlife biologists, in part to review data gathered and analysis done by state biologists and to compile accurate resource inventories. Their scientists were expected to help develop regulations that would allow tribal members to harvest off-reservation resources in ways that protected the resources, that were culturally appropriate and that might include methods not allowed by the state. The commissioners also recognized that wardens were needed to enforce laws and that wardens should be placed on each reservation. Finally, tribal court systems were supported with funding for operations and training.

Cooperative Management Defined

Cooperative management, is sometimes referred to by other terms such as coordinated, joint, or shared management. Regardless of what it's called, you know you have it if there is a commitment to work together, to share information, to communicate, and to reach decisions through consensus. Also, regardless of whether one government is viewed as the primary manager or whether power is shared, each government must act as a manager because each has its own right and its own responsibility for decisions and actions that affect shared resources.

The concept of cooperative management is not new. Gulland (1974) described a two-step process, developed over many decades, to manage international fisheries. First, fishery biologists representing all parties develop facts on the fishery and biological limiting factors; then, policymakers from the parties meet to develop a politically acceptable consensus. In recent years, cooperative management has generally stemmed from concerns over the perceived, and sometimes real, mismanagement of natural resources and the depletion of natural resources (Pinkerton 1989).

Tribal-state cooperative management arrangements are varied and each seems to have unique features. For example, the Great Lakes Indian Fish and Wildlife Commission (GLIFWC) is composed of 11 member tribes. These tribes have reservations in three states, and each state is composed of two different treaty areas with both inland and Lake Superior rights. If you multiply these factors out (three states times two treaties times two areas), you have 12 possible arrangements. Also, in two instances, GLIFWC member tribes have treaty rights in common with tribes not part of GLIFWC and have formed a different intertribal natural resource commission (i.e., Bay Mills belongs to the Chippewa/Ottawa Resource Authority [CORA] in Michigan).

Cooperative Management on Lake Superior: The Great Lakes Fishery Commission

In Lake Superior, cooperative fishery management exists among nearly all the governments with fishery responsibilities, including U.S. tribal, state, federal, and provincial governments; the only ones not included in this partnership are tribal governments in Canada. This cooperative management exists under the auspices of the Great Lakes Fishery Commission (GLFC). The GLFC was formed in the mid-1950s, primarily in response to the crash of the Great Lakes fish stocks due to overfishing and sea lamprey *Petromyzon marinus* predation.

The founding document of the GLFC that sets out the institutional arrangements and rules for making decisions is called the *Strategic Great Lakes Fisheries Management Plan* (SGLFMP) (GLFC 1997). The strategic plan is a nonbinding agreement by which agencies have "committed to the following: (1) consensus on management actions with interjurisdictional implications, (2) accountability for performance, (3) communication of environmental management needs, and (4) cooperation in measuring and predicting effects of fishery- and environmental management needs" (Dochoda and Koonce 1994).

Each of the upper Great Lakes has a lake committee composed of state, provincial, and tribal representatives. Lake committees develop politically acceptable consensus on fishery issues such as sea lamprey control, lake trout *Salvelinus namaycush* stocking, and fish community objectives. Lake committees are responsible for development of a *State of the Lake* and a *Fish Community Objectives* report and to review them every 5 years. A technical committee serves each lake committee by developing biological facts in response to assignments. Issues or policies that affect more than one lake are reviewed by the Council of Lake Committees. The ultimate decision-making body is the Council of Great Lakes Fishery Agencies.

Both GLIFWC and CORA have seats on the Lake Superior Committee (LSC) and the Lake Superior Technical Committee (LSTC). A recent accomplishment of these committees was to develop a second version of the document *Fish Community Objectives for Lake Superior* (Horns et al. 2003) This plan outlines common goals for restoring and sustaining the fishes of Lake Superior.

The stocking of nonnative salmonids, like Chinook salmon *Oncorhynchus tshawytscha* and coho salmo *O. kisutch*, in Lake Superior is an issue that agencies disagree on but carefully crafted language in SGLFMP has perhaps served to prevent this from becoming a dispute. One of the common goal statements for Great Lakes fishery agencies is "to secure fish communities, based on foundations of stable self-sustaining stocks, supplemented by <u>judicious</u> plantings of hatchery-reared fish ..." (underline added; GLFC 1997). Several tribal governments on the west end of Lake Superior are concerned about the potential negative effect nonnative salmonids might be having on cisco (also known as lake herring) *Coregonus artedi* and lake trout *Salvelinus namaycush*. Still, other Lake Superior agencies stock or support the stocking of these salmonids.

Cooperative Management on Lake Superior: Wisconsin—Tribal Negotiations

Independent of the existing GLFC institutional arrangements, two Wisconsin tribal governments and the state of Wisconsin have negotiated terms for implementing off-reservation fishing in Wisconsin waters of Lake Superior. The first agreement in 1981 only included the Red Cliff Band; in 1984, the Bad River band joined the process (Busiahn 1989). A state/tribal technical committee consisting of fishery biologists from the state and the tribes was formed to provide facts, advice, and assist negotiators in resolving biological disputes.

Quite often, when developing facts and advice, fishery biologists are able to rely upon documents produced, agreed to, or used by the Lake Superior Committee. For example, disputes about total allowable catches (TACs) can be resolved by turning to the lake trout catch-age models supported by the LSTC to determine TACs in U.S. management units, including those in Wisconsin. Questions about whether harvest levels are interfering with lake trout rehabilitation can be resolved by the fact that the

LSC and LSTC have agreed that rehabilitation can occur if total annual mortality is kept at or below 45%.

In some cases, tribes may choose to conduct independent studies to develop facts. For example, the Red Cliff tribe felt that fishermen should be allowed to release live lake trout but the negotiated agreement did not allow this. By conducting a study to determine survival of lake trout caught in gill nets during various seasons, facts were developed that led to an adjustment in an effort formula. This adjustment provided more opportunity for tribal fishermen to fish for lake whitefish *Coregonus clupeaformis*, the target species. Also, the Red Cliff fisheries biologist began to systematically collect lake trout diet information several years before the LSTC decided to.

This cooperative management arrangement within the 1842 area of Lake Superior is working. Signs of this are agreements are being reached, and as stated above, lake trout have been declared rehabilitated in Lake Superior, including waters from the Apostle Islands region of Wisconsin to central Michigan.

Cooperative Management on Wisconsin Inland Lakes

The 1983 Voigt Decision, which reaffirmed treaty-reserved rights in northern Wisconsin, was followed by a series of trials that determined, in part, that tribal regulations were either adequate or could be made adequate to protect the resources and that resources should be split 50:50. These court decisions, including stipulations, provide a foundation for cooperative arrangements to manage the fishery, wildlife, and wild plant resources in northern Wisconsin.

In 1984, an inland fisheries technical working group (TWG) was formed by the parties to develop biological facts on assigned issues. Subsequent stipulations and a court decision in 1989 institutionalized the TWG and gave it various duties. The parties also entered into a memorandum of understanding (MOU) "to provide for the timely and accurate exchange of routine fisheries information, provide for routine communication...for special information requests, (and) establish procedures for conflict resolution..." (Memorandum of understanding on exchange of information and communications between the Wisconsin Department of Natural Resources and the Great Lakes Indian Fish and Wildlife Commission, version 2, October 1994). The TWG was charged with meeting at least twice per year to coordinate sampling and survey schedules, to exchange and discuss routine fisheries information, and to make joint recommendations concerning mixed treaty-state fisheries.

"...to provide for the timely and accurate exchange of routine fisheries information, provide for routine communication...for special information requests, (and) establish procedures for conflict resolution...."

Despite the court decisions and the institutional arrangements to coordinate work and share information, an enormous amount of social unrest and controversy had developed from 1985 to 1989 about spearing walleye *Sander vitreus* during spring spawning. In 1990, a joint federal/tribal/state committee known as the Joint Assessment Steering Committee was formed to answer questions about whether tribal, off-reservation treaty harvest was depleting walleye populations in northern Wisconsin and about the status of the walleye resource. In a 1991 report, *Casting Light Upon the Waters*, the committee answered the questions at hand: "NO! - Chippewa spearing has not harmed the resource; and YES! - the fish population in the ceded territory is healthy" (U.S. Department of the Interior 1991) Funds provided by Senator Innouye gave GLIFWC and the tribes an opportunity to expand their assessment capabilities. Together with information collected by an increased WDNR assessment program, a comprehensive database on walleye in northern Wisconsin lakes was begun and added to each year. A 1995 follow-up report, *Fisheries Status Update in the Wisconsin Treaty Ceded Waters* (U.S. Department of the Interior 1995), provided the public with findings and an overview of the joint effort.

A health issue that became important as tribal harvest increased to around 30,000 walleye each spring was a concern about mercury in fish. GLIFWC biologists reviewed WDNR data and began collecting samples in untested lakes where tribal members were harvesting fish. In 1996, a second MOU (Memorandum of understanding on exchange of information and communications for fish contaminant testing with an emphasis on mercury between the State of Wisconsin Department of Natural Resources and the Great Lakes Indian Fish and Wildlife Commission, March 14, 1996 version) and was signed in which both agencies agreed to use each other's data, to coordinate sampling plans, and to strive for accuracy in assessing health risks to those who consume fish. Between 1996 and 2000, GLIFWC tested nearly 1,500 walleye fillets for mercury levels. Using this information, easy-to-read geographical information system maps were developed before the 2000 spearing season, which show tribal leaders, spearers, and their families where mercury levels may be of concern and where walleye are safe to eat.

"...institutional arrangements that provide for consensus decisions, for similar procedures to change regulations, and for managers to learn how to manage the fishery and harvest opportunity better could be important features to ensure fairness."

One issue that continues to concern Wisconsin tribes is that the current system for setting walleye quotas in inland lakes does not provide for a clear 50:50 allocation of the resources. Tribal biologists have been charged to work either independently or with state biologists to come up a fair and explicit system to split the resource 50:50. While seemingly simple, this task is made difficult by the fact that a separate system was developed by the state and adopted by the Court for determining the number of walleye tribes could take. This system for setting Safe Harvest levels, has built in barriers (e.g. risk levels, safety factors) that theoretically cap tribal harvest at one-third of the harvestable surplus. This system, including it's level of risk, does not apply equally to state anglers.

Tribes should be alert to avoid state systems that are created especially for their fisheries and apply only to them. Rather they should expect that systems for sharing resources 50:50 be fair and that only one method be used to develop a harvestable surplus value (i.e., TAC). Also, institutional arrangements that provide for consensus decisions, for similar procedures to change regulations, and for managers to learn how to manage the fishery and harvest opportunity better could be important features to ensure fairness.

References

Benton-Banai, E. 1979. The Mishomis book: the voice of the Ojibway. Red School House, St. Paul, Minnesota.

Busiahn, T. R. 1989. The development of state/tribal co-management of Wisconsin fisheries in cooperative management of local fisheries. Pages 170–185 in E. Pinkerton, editor. Cooperative management of local fisheries: new directions for improved management and community development. University of British Columbia Press, Vancouver.

Dochoda, M. R., and J. F. Koonce. 1994. A perspective on progress and challenges under a joint strategic plan for management of Great Lakes fisheries. University of Toledo Law Review 25(2):425–441.

GLFC (Great Lakes Fishery Commission). 1997. A joint strategic plan for management of Great Lakes Fisheries. Great Lakes Fishery Commission, Ann Arbor, Michigan.

GLIFWC (Great Lakes Indian Fish and Wildlife Commission). 1992. Wii Gimawanjii'idimin Gaye Wii Nibawaadaanamin (We will all meet together and we will dream). Strategic directions for the Great Lakes Indian Fish and Wildlife Commission in off-reservation treaty resource management. Great Lakes Indian Fish and Wildlife Commission, Odanah, Wisconsin.

Gulland, J. H. 1974. The management of marine fisheries. University of Washington Press, Seattle.

Horns, W. H., C. R. Bronte, T. R. Busiahn, M. P. Eberner, R. L. Eshenroder, T. Gorenflo, N. Kmiecik, W.

Mattes, J. W. Peck, M. Petzold, and D. R. Schreiner. 2003. Fish-community objectives for Lake Superior. Great Lakes Fisheries Committee, Special Publication 03-0, Ann Arbor, Michigan.

Pinkerton, E., editor. 1989. Cooperative management of local fisheries. University of British Columbia Press, Vancouver.

U.S. Department of the Interior. 1991. Casting light upon the waters: a joint fishery assessment of the Wisconsin ceded territory. U.S. Department of the Interior, Bureau of Indian Affairs, Minneapolis, Minnesota.

U.S. Department of the Interior. 1995. Fisheries status update in the Wisconsin treaty ceded waters. U.S. Department of the Interior, Bureau of Indian Affairs, Minneapolis, Minnesota.

American Fisheries Society Symposium 52:169–171, 2006

Cooperative Research in the Great Lakes: Exploring Characteristics of Success

ROY A. STEIN
Aquatic Ecology Laboratory
Department of Evolution, Ecology, and Organismal Biology
The Ohio State University
1314 Kinnear Road. Columbus, Ohio 43212, USA

CHARLES C. KRUEGER
Great Lakes Fishery Commission
2100 Commonwealth Boulevard, Suite 100
Ann Arbor, Michigan 48105, USA

Introduction

Our paper describes the unique nature of cooperative fishery research and management in the Great Lakes and formulates operational procedures based on our experiences that may be useful elsewhere.

Cooperative Great Lakes Fishery Research

Cooperative fishery research in the Great Lakes basin occurs at various physiogeographical, organizational, and political scales. For example, at one end of the spectrum, cooperative research programs occur, as described below, between a state fishery management agency and a university on a single lake, Lake Erie. Basin-wide, international coordination, focused on a specific research theme and coordinated by the binational Great Lakes Fishery Commission, represents the other end of the spectrum. Below, we discuss these two examples. Our first example focuses on a long-term relationship between the Ohio Division of Wildlife (ODOW) and the Ohio State University on Lake Erie. A second example describes how resources and projects are coordinated across the Great Lakes basin to understand the role of exotic invertebrates in disrupting Great Lakes food webs.

Ohio Division of Wildlife and Ohio State University

In 1978, and with leadership from the Ohio Cooperative Fish and Wildlife Research Unit, a research partnership between Ohio State University and the ODOW was established. Through time, the partnership has grown, providing substantive funding to university research programs, supporting graduate students and faculty, and leading to science-based management, both on inland reservoirs and Lake Erie. During the past 15 years or so, the partnership truly has flourished as the university (and its associated infrastructure, funded in part by ODOW) grew into its current role as the division's research, experimentally oriented branch, working cooperatively and collaboratively to solve outstanding fish management problems within the state. Operationally, the university benefits via direct funding of research projects, secretarial support, research equipment, building and equipment support and maintenance (including computers), travel to scientific meetings, training, and workshops. In turn, the partnership provides opportunities for (1) internships for graduate students, (2) use of historical data sets (from Lake Erie and our inland reservoirs), (3)

“Cooperative fishery research in the Great Lakes basin occurs at various physiogeographical, organizational, and political scales.”

collection of field data by ODOW personnel in direct support of research projects, and (4) in-kind support of research.

Unique to this partnership is how graduate students decide upon their thesis research. Typically, the ODOW approaches the university with a topic of concern, such as the impact of the invasive round goby *Neogobius melanostomus* on smallmouth bass *Micropterus dolomieu* in Lake Erie. Once this topic is broadly identified, graduate students are recruited, then required to meet with division biologists for the purpose of defining research goals and objectives. Via multiple meetings during their first fall of residence, graduate students present and defend (with typically strong direction from their major adviser) their perspective on the problem. Through these meetings, consensus is reached on the goals and objectives of the research effort, as well as the student's thesis. In our view, this collaborative, agency–university interaction serves both the division and the student.

Research emanating from this collaboration has been immediately applied to management. For example, work on smallmouth bass–round goby interactions in the Bass Islands of the west basin of Lake Erie led to a regulation change resulting in a spring angling closure. As another example, ongoing work with microelemental analysis of walleye *Sander vitreus* otoliths will have important implications for determining the distribution of walleye production across reef- and river-spawning populations, allowing managers to differentially protect stocks based on their contribution to the walleye fishery in Lake Erie. This partnership is a two-way street: students gain insight into agency work while the division profits from the research findings.

Establishing this partnership was driven by visionary individuals within the division who believed their management would be improved by rigorous scientific research. That this perspective would fall on fertile ground within the university was significant and well facilitated by flexible and broad-thinking administrators. Critical to this unique partnership was the willingness of both parties to adapt their respective and sometimes rigid organizations to bring a productive program to fruition.

Food–web disruption research

The most serious challenge facing Great Lakes' fisheries is the potential catastrophic alteration of aquatic food webs by invasions of Eurasian nonnative species. Both vertebrate (e.g., round goby) and invertebrate (e.g., zebra mussels *Dreissena polymorpha*) invaders successfully have colonized the Great Lakes in the last 20 years, forever changing fisheries and their supporting food webs. Coordinated, cooperative research on these effects has been needed to provide fishery management agencies with the information to readjust fish community goals in light of new ecological realities.

In 2001, the Great Lakes Fishery Commission formed a partnership with eight other research-funding organizations, the Great Lakes Fishery Trust and the seven Sea Grant programs in the Great Lakes basin, to focus funding on understanding how four nonnative invertebrates, zebra mussels, quagga mussels *D. bugensis*, spiny water flea *Bythotrephes longimanus*, and fishhook waterflea *Cercopagis pengoi*,S influence food webs of the Great Lakes. This partnership sought to support studies that will enable fishery managers to minimize the impacts of altered food webs on their fisheries. The research theme is described in a white paper entitled "Exotic Invertebrates, Food-Web Disruption, and Lost Fish Production: Understanding Impacts of Dreissenid and Cladoceran Invaders on Lower-Lakes Fish Communities and Forecasting Invasion Impacts on Upper-Lakes Fish Communities" (http://www.foodwebdisruption.org/) by Brian Shuter (Ontario Ministry of Natural Resources) and Doran Mason (NOAA-Great Lakes Environmental Research Laboratory). Within this research theme, the state of knowledge, ecological

"Critical to this unique partnership was the willingness of both parties to adapt their respective and sometimes rigid organizations to bring a productive program to fruition."

rationale, and key research questions are developed. About 25 research projects have been funded by the partnership during the past 5 years to investigators in federal, provincial, state, and university organizations. Communications among the researchers also has been a high priority. Several workshops for partnership-funded researchers and others have been held, mostly in conjunction with the annual meetings of the International Association of Great Lakes Research.

Several advantages of cooperative research exist within the context of the Great Lakes. First, understanding the role of nonnative invertebrates on Great Lakes food webs requires funding scales in excess of any single organization. Through this partnership, substantial financial resources can be focused on this problem in a coordinated manner. Second, by including several partners, the research problem can be addressed at an appropriate spatial scale. For example, the Great Lakes Fishery Trust focuses on Lake Michigan, whereas Minnesota Sea Grant concentrates on Lake Superior. By working through the Great Lakes Fishery Commission, the combined resources of all partners can be brought to bear across all five Great Lakes in both Canada and the United States. Finally, the food-web disruption theme provides guidance, ensuring that funded projects are consistent with a common rationale and set of priorities. Similar to cooperative management, success of this research partnership depends on key people within the research funding organizations. Because research administrators and their organizations were open to new ideas and new ways of conducting business, this partnership became a reality.

“The Great Lakes have an organizational structure where all management agencies have separate but equal voices, where stakeholders are advisory, where decisions are made by consensus, and where an oversight body with credibility (but no management authority) serves to arbitrate difficult issues.”

Applying the Great Lakes Experience

The Great Lakes have an organizational structure where all management agencies have separate but equal voices, where stakeholders are advisory, where decisions are made by consensus, and where an oversight body with credibility (but no management authority) serves to arbitrate difficult issues. Because the system is voluntary, protects and respects individual jurisdictional authority, and requires written goals and objectives, it works. Finally, it works because strong, though cooperative, charismatic leadership characterizes the managers within the basin. Their desire for consistent ecosystem management overrides local, more restricted perspectives, which are, at times, embraced by their administrators.

Cooperative research, in which agencies and universities forge partnerships, and in which funding agencies join forces to attack basin-wide problems, also is driven by effective organizational structures, but most importantly by flexible, creative administrators. Within the Great Lakes, we have been indeed fortunate to have innovative administrators that serve to deal effectively with ever-evolving ecosystem issues. Our view of cooperative management and research and its attendant difficulties lies, first and foremost, with people … success requires fishery managers that have courage and credibility … critical character traits

Opportunities

American Fisheries Society Symposium 52:175–178, 2006

Cooperative Research: Needs and Interests

CLARENCE PAUTZKE
North Pacific Research Board
1007 West 3rd Avenue, Suite 100, Anchorage, Alaska 99501, USA

Introduction

In helping to draft the science plan for the North Pacific Research Board in 2004, I wanted to include a section on cooperative research. I found that the best guides to the overall pros and cons of using cooperative research was the National Research Council (NRC) report titled *Cooperative Research in the National Marine Fisheries Service*, published in 2004, which is the basis for this discussion. I intend to talk also about cooperative management at the September 2005 American Fisheries Society conference, but this piece focuses on research.

" I have observed that there are many very sophisticated fishermen out there who closely observe the marine ecosystem they work in and often provide fisheries and marine mammal-related data and analysis in their comments that rival the scientific information coming from the agencies. "

I heartily agree with the NRC's first finding in its report that most fisheries research projects would benefit from some level of cooperation with industry. Fishermen in general are very knowledgeable of fishing gear, fishing grounds, and fish behavior, and this knowledge can be incorporated in most forms of research and the formation of hypotheses to be tested. We also need to face the basic fact that funding for marine research is in short supply and ship time is expensive. Most fishermen clock a lot more days on the grounds than most researchers who have to pay for ship time and may only spend 5–10 d on a cruise each year.

Having sat through many hours of testimony at North Pacific Fishery Management Council meetings over 20 years, I have observed that there are many very sophisticated fishermen out there who closely observe the marine ecosystem they work in and often provide fisheries and marine mammal-related data and analysis in their comments that rival the scientific information coming from the agencies. Indeed, commercial fishermen offer a significant opportunity to collect scientific information on the fisheries and marine ecosystem. They provide field experience, practical knowledge, and platforms for collection of data. They are expert at deploying their gear and have the knowledge to increase their efficiencies and lessen their impact on non-target fish stocks. Their expertise can be important in making sure that survey fishing gear is operated as efficiently as designed or that the geographic range of the survey is consistent with the geographic range of the fish. Use of fishing gear in research helps scientists better understand the impacts of that gear, not only on the fished population, but also on the surrounding habitat. It also allows intercalibration of gears used by scientists and fishermen.

In contrast, scientists are more likely to be the ones that can contribute experimental design, the scientific method, and data synthesis. By bringing together the knowledge and skills of these two groups, the quality, quantity, and relevance of research may be improved. Working together may help to build a better understanding between science and industry and greater confidence in the products of research and in the regulatory process. Cooperative research efforts must ensure the scientific integrity, practicality, and cost effectiveness of the experimental design and facilitate the ready application of the results to alter fishery management if the results suggest that such alteration is beneficial or required. Cost effectiveness, practicality, acceptability, and utility must be key design criteria.

Forms of Cooperative Research

Cooperative research may take a variety of forms:

1. Industry may assess a tax on their landings and directly fund research, possibly in partnership with other entities.
2. Industry may provide vessels and crew for charter by research scientists, which may reduce costs of research and augment ship time available on federal research vessels. The vessel crew adds their expertise, including seamanship, fish finding and gear handling. The International Pacific Halibut Commission and National Marine Fisheries Service charter vessels to survey halibut and sablefish *Anoplopoma fimberia*, respectively.
3. Industry may participate in experiments to modify gear to reduce bycatch. Participation of the North Pacific Longline Association and the Fishing Vessel Owners Association in seabird deterrent research, Groundfish Forum work on bycatch reduction gear through exempted fishing permits, and NPRB-funded research on salmon bycatch reduction, all are good examples of this cooperative approach. The direct cooperation between industry and government or university scientists provides a realistic setting for testing gear modifications, reducing costs, and making the results more believable to industry. Parties work together to formulate, secure funding for, execute, and analyze a project. Either the industry or the scientists may initiate the project.
4. A group of fishermen may get together to do a study or survey, consulting a scientist as needed to ensure scientific validity.
5. Fishermen can maintain logbooks to provide information on catch, bycatch, and other aspects of the fishery. They also can retrieve tags from tagging studies.
6. Vessels can and do take observers that provide a variety of catch and other information. For example, all vessels greater than 60 ft and fishing groundfish off Alaska are required to have one or more observers on board.
7. Fishing (and other commercial ships of opportunity) vessels can take scientific data collection instrumentation packages such as continuous plankton recorders, ferry box systems, or other oceanographic data packages to collect information over a wide region.
8. Fishermen can contract with outside scientists with no affiliation to the management agency that governs their fishery. One purpose for this can be to improve on methods or assumptions affecting stock assessments or other scientific reports relevant to fishery management decisions.

Information on each cooperative project funded by the North Pacific Research Board over the past 4 years is available at www.nprb.org (go to research links).

Elements of Success

I believe that the most successful and popular cooperative research partnerships cluster around projects where fishermen's vessels, gear, and expertise can be readily employed and where there is the strong potential for research results to substantially change assessments and regulations to provide short- and long-term economic benefits to fishermen. Other substantial objectives include improving stakeholder trust in the fishery science and management system, improving research methods and administration, and coeducating scientists, industry, and managers. Cooperative research tends to work best when scientists and fishermen acknowledge early on that a problem needs to be addressed and that working together cooperatively will be the most effective way of proceeding. They must realize that each bring

> "...use each other's expertise in science and on the fishing grounds to design the most appropriate and practical research protocol."

valuable tools and experience to the objectives of a research project and that they should use each other's expertise in science and on the fishing grounds to design the most appropriate and practical research protocol. Participants must understand that they are involved in scientific research and must meet scientific standards for the results to be credible and useful for management decisions. Finally, they must properly interpret and timely distribute the results of their project to affected parties and management agencies.

In summary, elements of successful cooperative research projects include

1. substantial incentives and benefits to research partners.
2. rigorous coengagement in most elements of the scientific process.
3. complementary skills and abilities.
4. honesty, trust, and mutual respect.
5. adequate financial, administrative and scientific support.
6. ongoing funding sources and, in some cases, an institutional framework dedicated to cooperative research that helps in setting long-term objectives, structuring partnerships, developing efficiencies in proposal development and evaluation, and coordinating and developing long-term partnerships where appropriate and desirable.

The North Pacific Research Board uses a competitive review process to allocate cooperative research funds, and the resulting data, analyses, and reports must undergo peer review. We need commercial fishing vessels used for cooperative research to meet all U.S. Coast Guard requirements for operation and manning to ensure safe operations and to have all required federal permits. There should be appropriate liability insurance to protect the financial interests of all participants involved in cooperative research. Agreements and contracts should include project design, project implementation, contingency plans, data collection analysis, and communication responsibilities and decision-making authority. And last, there should be a communications plan for outreach, progress reports, and dissemination of final results, while respecting the confidentiality and ownership of some forms of data and information. Rules governing vessel charter arrangements and exempted fishing permits often affect confidentiality of data, the degree to which gear innovations and technologies developed in the research can remain proprietary, and the disposition (including sale if allowed) of fish harvested during field work. Recognizing these limitations, care should be taken from the outset to select a cooperative research vehicle that meets the short- and long-term expectations for all parties in the research.

> "Participants must understand that they are involved in scientific research and must meet scientific standards for the results to be credible and useful for management decisions."

Implementation Strategies off Alaska

The North Pacific Research Board has the following strategies identified in its science plan for implementing a cooperative research program off Alaska, and they will be considered over the coming years:

1. Seek opportunities to partner with industry-funded research entities.
2. Identify ships of opportunity to collect oceanographic and other data over a wide region.
3. Establish a registry of commercial vessels potentially available for research charters.
4. Support development and deployment of cost-effective instrumentation packages.
5. Use commercial vessels as a cost-effective way to do ecosystem monitoring.
6. Convene workshops with industry and scientists to identify research needs, develop hypotheses, and establish ways to partner.

7. Encourage conservation engineering projects (e.g., bycatch reduction research) in Request for Proposal.
8. Sponsor training in fisheries science and collection of scientific data to enable fishermen to participate in all facets of fisheries research, from identifying research questions to developing project procedures and protocols, collecting data, and peer review of results.
9. Help establish for Alaska a central cooperative fisheries Web site such as www.fishresearch.org.
10. Establish a formal cooperative research program to develop long-term funding sources, develop cooperative partnerships and relevant expertise, and foster development of cooperative research where appropriate.

Summary

In summary, the North Pacific Research Board has already invested well over a million dollars in cooperative research-related projects, and I expect the program to grow robustly over the coming years. I believe we will be very responsive to Recommendation 19-9 of the U.S. Ocean Commission Report, which calls on NOAA Fisheries to create an expanded, regionally based cooperative research program that coordinates and funds collaborative projects between scientists and commercial, tribal, and recreational fishermen. For all the reasons given above, cooperative research activities should be included in any comprehensive science program, particularly those that influence how living marine resources are managed. Commercial fishermen, working with marine fisheries scientists, can provide a rich source of information critical to sustainable management.

“Commercial fishermen, working with marine fisheries scientists, can provide a rich source of information critical to sustainable management.”

American Fisheries Society Symposium 52:179–184, 2006

Session 3A: Opportunities

Linking Cooperative Research and Management: Partner's Needs and Interests

Alesia N. Read
American Fisheries Society
5410 Grosvenor Lane, Suite 110, Bethesda, Maryland 20814, USA

Cooperative management and research roll right off the tongue, but the terms conjure up very different images for me. When I think of the phrase cooperative research I have positive images of sweetness and light: fishermen and scientists happily working together and maybe even whistling while they work, having a beer together when the day is done, puffy white clouds, blue skies and calm weather—that sort of thing. In contrast, when I think of cooperative management, I think of tension and fighting, turbulence, gray clouds, low pressure systems, people hunkering down as they hammer out management regimes worth millions of dollars to various sectors of the fishery. [Clarence Pautzke, executive director of the North Pacific Research Board]

John Schwartz, director of Michigan Sea Grant Extension located in the Department of Fisheries and Wildlife at Michigan State University moderated the panel and audience discussion.

Keynote speaker Clarence Pautzke, executive director of the North Pacific Research Board, set the stage for the panelists and discussion on how we can begin to integrate cooperative research and management.

Clarence Pautzke, executive director of the North Pacific Research Board

- Regional fishery management councils (established in 1976) are a prime example of cooperative management and are good models for study.
- There are few instances of complete agreement on how to manage fisheries; disagreement between opposing factions will increase the more there is at stake in terms of allocation and wealth.
- "Scientists in the world of fisheries management have very difficult jobs; there is tremendous pressure to provide accurate and defensible science. Scientists that will succeed in this environment are those that are cool under pressure and doubly confident of their information. Your analysis and conclusions will end up in analytical documents and in court proceedings, you may be consulted by industry; so your information needs to be as scientifically sound and irrefutable as you can make it."
- In the rush to embrace cooperative research, we must squarely keep in mind that the end game is not the cooperative research. The end game is the scientifically sound information.

- "It is important to be careful in defining information needs before defining cooperative research needs. What are the issues in the geographic area? What is the scope of the project? How many millions of dollars are on the table? Are there different interested users with different gears? How many different regions are involved? Will one group believe another group's research? The devil is in the details; a robust experimental design is necessary to underpin the eventual conclusions."
- It is essential that partners provide information back to everyone who was involved with the project. There is a tendency for researchers to do research and get the paper published, and they never go back out and close the loop with the industry, the community, or the people who could really use the information.

Recurring Themes and Emphasized Ideas

Several themes emerged amongst the panel and audience discussions. The questions that were put forth for this discussion session were focused on partners' needs and interests, topics that are ripe for more research, and the current fisheries governance structure. Several key issues surfaced and are discussed in the following paragraphs.

Appropriate matching of cooperative partners

Several panelists mentioned the importance of proper and appropriate pairings of cooperative partners. Gary Graham shared a situation where a successful cooperative workshop was completed and

Chris Glass, Manomet Conservation Science Center, Massachusetts

- Each specific program has different needs and interests; in some cases, they are significantly different needs. Any perception that one partner's needs are being met at the expense of another partner will weaken the program, and potentially render it ineffective.
- Key is communication, clearly articulate all of the goals *at the start of the cooperative program*, address all needs, voice concerns, establish clear measures of success; buy-in is necessary from the onset.
- "We need to be very careful about setting narrow bounds around cooperative programs; what I mean is that many cooperative programs arise because of the urgent need for data in support of a particular management issue and/or action. But broader scientific initiatives might be necessary, crucial components might be large and broad and will not be available on an urgent basis, ex. You might need a larger time scale of data."

Gary Graham, Texas Sea Grant

- Over time, fisheries management has gone through many changes.
- At first, there were the dark days in cooperative research—passing Magnuson-StevensFishery Conservation and Management Act opened up a lot of opportunities and tons of boats entered the fishery. The buzzword was "underutilized species," which created a rush to develop lots of fisheries. There was no managing structure to handle this growth in the fishery. The industry was very successful at harvesting fish, but without a management structure in place to regulate the fishery, the result was widespread depletion of the resources.
- There was a shift from the dark years to the contentious years in cooperative research—mandates came down and there were lots of problems.
- Interestingly, sometimes the most complex problems have easy solutions—for example, turtle excluder devices are an example of industry-developed gear. Industry-established gear changes work well—the fishermen know they work efficiently, and there is buy-in for their use.
- Presently, we are in a stage of reduction devices—through ownership and industry developing gears; cooperative research has become increasingly successful.

Sean Martin, commercial fisherman, Hawaii

- A bottom-up approach is best, where fishers are participating in research and adopting methods through experiments and demonstrations; they are directly involved in developing and testing the gears.
- When industry ownership is fostered, compliance with the rules and regulations increases.
- Litigation is not an efficient use of time or money. Collaborative research and finding solutions to problems together is cheaper than litigation.

Chuck Krueger, Great Lakes Fishery Commission

- Needs of managers: Jurisdictional respect; managers are responsible for fisheries and are held accountable by the public. The recognition of public trust is necessary the responsibility of a manager is to serve the public and also to serve all stakeholder groups, even those who have no voice in management: children, subsistence users, and future generations. It is not simple to be a good manager; they need to be provided with objective (unbiased) ecological and scientific information, as well as information on the uncertainty associated with these data.
- Needs of fishermen: Understandable information that does not contain specialized vocabulary; use of this language gives the appearance of superiority and it is not clear. They need to have a clear description of the variation in the data and be given an idea of what the uncertainty is like. They need open channels of communication; nothing is more frustrating than to be made to feel like your opinion or information does not matter when you have spent years on the water and have lots of experience.
- Needs of scientists: Need a clear identification and description of what type of information will be helpful and valuable for managers; this will help them to set priorities and research goals in their research programs.
- Real cooperation means full participation at the decision-making table; this means full vulnerability and an acceptance of responsibility to being held accountable to the consequences of the decisions that were made together.

the attendees tried to take the lessons they learned out to their respective regions and put the tools into practice. This did not work, and he attributed this to the partners involved. Certain people work well together, but partners need to be chosen carefully. Each cooperative program is different; cultural and regional differences need to be taken into account. There is not a blanket or a one-size-fits-all procedure that will work for cooperative projects.

People with a predisposition for participation seem to work well together. Cooperation depends on the people, first and foremost. The essential elements for a successful program include trust, honesty, fairness, communication, and relationship development—this will take time and effort. One panelist noted, "The best governance structure will fail without these elements. However, cooperation can occur when these elements are present, but there is no governance structure."

There is great power in giving people real responsibility. Industry-led, industry-developed research projects that put the responsibility of managing the resource on the fishermen have proven to be effective. Working from the ground up with full and equal participation from the onset of a project leads to successful cooperative programs, increased support for cooperative programs, increased buy-in of the results, and most importantly, increased understanding of how and why science and management are structured as they are. This also opens up dialog into innovative changes that can be made to create a more efficient management structure.

Need for high quality science

All science must be held to the same rigorous standards. In order to make informed management decisions, all science must be the best available, rigorous, peer-reviewed science. Symposium participants discussed the perception that cooperative research has not yet proven its rigor or might not be the highest quality science and were glad to explore where this might be rooted so as to resolve this misconception. The panel agreed that regardless of who is participating or conducting research, everyone must use scientific tools correctly, appropriately, and consistently. Cooperative fisheries research can and does meet this standard. We should be concerned with the quality of all science, not just cooperative fisheries research. Science should be judged on its merit and not on those who produced it.

Cooperative participants stressed that they have observed much value added from the inclusion of all stakeholder parties throughout the research process: from the formation of hypotheses to the analyses of data collected. Broader perspectives brought to the cooperative research "table" yield products with more effective management implications and results that inform wider reaching interested groups. Involvement of all partners from the start increases the quality of the end result, but also empowers partners and increases buy-in and therefore compliance with rules and regulations. It is important to make sure the most capable individuals are involved, use a rigorous scientific design, and apply scientific tools in an objective manner. Clarence Pautzke suggested that we "...look at recommendation 19-9 of the Ocean Commission, it urges us to look for a process of external evaluation for all cooperative research proposals to ensure the most worthwhile projects are funded, the most capable performers are undertaking the research, and the information produced is both scientifically credible and useful to managers."

There was consensus among the panelists and audience members that we should never do cooperative research just to cooperate; it is essential that the goals of the project, the hypotheses, and the experimental design are clearly defined.

Different partners, different regions, different needs

Panelists offered myriad advice explaining the different needs of cooperative partners: scientists, managers, industry, community, and the public. Consideration of each partner's needs is essential. It is important to be sensitive to unique regional and cultural differences. For example, a particular fisherman may be participating in a cooperative project, but while he/she is gathering data, his/her fishing days are also being used. These are days when they could be fishing and making money. Provisions for this help should be clearly defined (i.e., pay for boat time, fish sale provisions that defer costs without creating an incentive to fish while conducting research). Compensation and consideration of each partner's needs should be formally documented at the start of the project. This will build trust, ensure respect for everyone's needs, and prevent any future misunderstandings that could undermine project progress.

Close the information loop

Panelists also emphasized the need to close the information loop after completion of projects. Follow-up is essential for both the expansion of knowledge and for dissemination of information. One panelist urged, "Do not let reports gather dust. Share your findings with the public and most importantly with the partners that helped you do the work." Chris Glass shared that there are likely hundreds of projects where the results fulfill the grant requirements and the reports are piled on desks gathering dust. Programs that are successful are ones where "the scientists make sure that results are discussed in the right places, brought to the councils, and shared with the appropriate people." It is helpful to involve people who understand that the results of the project are important and are able to communicate this. Funding bodies should encourage partners to make follow-up items requisite in proposals; this will ensure that these items are carried out. Clarence Pautzke noted that it is helpful to "make outreach and education programs part of your overall research. Require partners to include money in their proposals for education and outreach, to create pamphlets and encourage travel to share the information gained."

Efficient use of time and money through cooperative programs

Chuck Krueger noted, "Much time is spent chasing dollars. The fact is, whoever pays the bills sets the research priorities. Funding bodies need to be more active participants in the conversation. Funders should contribute to research goals and expectations so that the right questions are being asked, but also so that they maintain the ability to be flexible enough so that the unforeseen solution can come out of the research." Throughout the symposium, it was stressed that cooperative research is the most efficient way to use resources, stop duplicating research, and spread knowledge.

It was also mentioned that use of litigation to set research or management objectives and priorities is not an efficient use of time or money. Opposing parties would increase their efficiency and productivity if they would cooperate. This is not always possible, and therefore, litigation is a necessary strategy to induce action; it should not, however, be the first option employed to settle a disagreement. A panelist posed the following question, "Litigation is a waste of valuable, already limited, resources. Cooperation is a more efficient problem-solving technique. How can we eliminate this step, and just go directly to the cooperative research without the involvement of the courts?"

The overall consensus was that court battles may waste time and money, but there also needs to be a deadline, crisis, or looming consequence to motivate and stimulate action. Oftentimes, the health of a fish stock or the preservation of valuable habitat may be at risk. Pressure of possible litigation provides motivation for action. Partners might expedite progress if they know their choice: "...either they figure things out or something will be figured out for them."

Topics ripe for cooperative research

Below are a few of the topics raised as potential fields that would benefit from increased cooperative programs:

- Long-term monitoring
- Habitat protection
- Ecosystem-based management
- Model development—model use for adaptive management
- Prediction of fishery and community response to regulation changes (markets, socioeconomic impacts, etc). Do the regulations meet the objectives they were meant to?
- Allocation of harvest and the involvement of all stakeholder groups
- Better stock assessments and mixed stock management

American Fisheries Society Symposium 52:185–187, 2006

Linking Cooperative Research and Management: Partners' Needs and Interests

Christopher W. Glass*
Manomet Center for Conservation Sciences
81 Stage Point Road, Post Office Box 1770
Manomet, Massachussets 02345, USA

Despite the plethora of recent cooperative research initiatives, cooperative research is not new. It has existed in many different forms—from single scientists making observations on individual commercial fishing vessels to larger more obvious collaborations involving multiple vessels and institutions—often without being recognized as such. However, the current, high-profile initiatives (in North America, in particular, but not exclusively so) are among the first to attempt to bring fishermen, scientists, managers, and regulators together in a meaningful and systematic manner. Such an integrated approach provides enormous potential for better and more effective management of fishery resources, but despite, and maybe even because of, the wide scope and ambition of some of these initiatives, such an approach inevitably also has inherent problems or challenges.

Partners' Needs and Interests

The major challenge presented by the formal linkage between managers, fishermen, and scientists is ensuring that the needs of all partners are adequately met. It is clear that each constituency will or may have different expectation of collaboration. Simplistically put, managers/regulators will be most interested in data and or studies that address specific management issues or that support specific management actions. Fishermen or the fishing industry may place more emphasis on economic and socioeconomic aspects of the collaboration, while scientists may inevitably focus on scientific goals of the program. The end result of the collaboration will have implications for all partners, but the specific needs may be met to a greater or lesser degree for individual partners. Perception that one partner's needs are being met at the expense of others may ultimately weaken the effectiveness of the cooperative venture and therefore the potential for effecting change. One very obvious example of this difference in needs being met can be seen in the field of bycatch reduction studies. There is a huge dichotomy between the need to (from a management and or scientific perspective) reduce capture of undersized fish and the need to reconcile this with fishermen's desire to catch all marketable sized fish. Clearly, any study that demonstrates complete reduction of bycatch and discard, but which releases a proportion of above minimum landing size fish, is likely to be considered successful by scientists and managers but may be considered less successful or even unacceptable to some or all in the fishing industry. The reverse is also true. Studies that show full retention of marketable fish with a small but significant bycatch and discard may be considered more acceptable to some fishermen than managers. This difference in perspective and expectation could be addressed by establishing clear measures of success

"...each constituency will or may have different expectations of collaboration."

* Current address: Northeast Consortium, Morse Hall, 39 College Road, University of New Hampshire, Durham, New Hampshire 03824, USA.

and by defining clear goals and objectives at the outset of cooperative research and management programs. In order that these programs are ultimately successful, these objectives, goals, and measures of success should be formulated by a collaboration of partners and should be subject to periodic review.

An associated challenge relates to the often narrow management needs imposed on cooperative programs. Most programs originate in a need for data in support of a particular management issue. This problem solving or reactive approach to cooperative research can restrict the scope of scientific thinking and may, in some cases, divert funding from longer term, more proactive initiatives. If science is to adequately support the needs of management and other partners in cooperative programs, it is imperative that a concerted effort be made to broaden the scope of experience and expertise in the scientific community and to promote broad scientific initiatives that are not continually focused on short-term management needs. For example, knowledge of fish behavior, migration patterns, and movements are often crucial components in addressing bycatch events and in understanding relative fish distributions and relationships between fish and associated habitat. However, gathering these data at short notice to address an urgent management dilemma (for example, a recent haddock bycatch event in the northeast region herring trawl fishery in 2005) is impractical at best and more usually impossible in the time scales involved. One way to address this need for more proactive research could be to make it appealing for more mainstream academic organizations (that generally have paid scant attention to fisheries issues) to turn their attention to aspects of the marine ecosystem and to promote the formation of scientific collaborations aimed at generating long time-series data-streams. In addition, to further increase the scope of scientific input, funding sources should be identified that are not coupled with management needs. This would provide impetus for researchers to address issues that may appear initially to be peripheral to addressing management needs but that, in the long term, may provide base data that enable us to better service requests from managers and regulators.

What Topics Are Ripe for Cooperative Research?

Two main areas of conventional research particularly lend themselves to a cooperative approach: (1) gear modification and or development of alternate fishing strategies to minimize bycatch and maximize target catch, and (2) real time studies on fish distribution in relation to environmental variables and habitat. Research in these broad fields has formed the core of cooperative research programs in the northeast region of the United States with a particular initial emphasis on gear technology and bycatch reduction research. These types of studies often offer quick solutions and can be used as a measure of initial success of a collaborative initiative. However, cooperative research programs continually evolve. Again taking the example of recent initiatives in the northeast region, programs have evolved over a short period of time (3–6 years) from being primarily problem-solving initiatives (e.g., gear technology/bycatch reduction research) to developing cooperative research projects that deliver long time-series data streams. For example, current, core cooperative research initiatives in the northeast region now include formation of study fleets, development of industry based surveys, and long-term studies relating to movement and migration of fish. This evolution away from a primary emphasis on problem solving has demonstrated the need for provision of basic information that can often be gathered on a daily basis, utilizing the fishing industry on a cost-effective basis. It is also becoming increasingly obvious that this basic data gathering can have a profound, long-term effect on management and stewardship of marine resources.

"...basic data gathering can have a profound, long-term effect on management and stewardship of marine resources."

Who Else Should Be Involved?

Cooperative research, for the most part, exists as a set of loose relationships between fishermen and scientists addressing questions relating to management needs. In other words, the research agenda is set by managers and regulators. In an advanced cooperative research and management process, the research agenda could, or perhaps should, be driven as much by scientists and fishermen as by managers and regulators. Furthermore, other stakeholders or others with an interest in promoting a healthy and sustainable marine environment should have a voice in helping set the agenda for cooperative programs. Without complete openness and opportunity for all, cooperative programs may be seen as self-serving, and this may greatly impact potential and longevity of such programs.

Collaborative research and management programs are now firmly established in parts of North America and overseas. The learning curve has been steep, but many of these initiatives are now beginning to impact management of our collective marine resources. Despite past successes, there is much to be achieved, and there is a real need to continue to refine our approaches to and philosophy surrounding collaborative research and management. It is the opinion of the author that the future of collaborative research lies in a vibrant, active (proactive perhaps?) fishing industry, collaborating with scientists and managers, collecting data while conducting fishing operations at sea, and making these data available to all interested parties perhaps on a real time basis. In order for this to be possible, it will be necessary to systematically develop infrastructure supporting such a process. This may include development and provision of instrumentation for use on fishing vessels, training for all parties involved, formation of a stable data repository, development of data processing and visualization systems, and education and outreach programs, to name but a few. In addition, and perhaps more importantly, there will need to be a systematic, well-formulated scientific structure underpinning and overseeing the data collection process that reflects the needs and expectations of all partners.

"Without complete openness and opportunity for all, cooperative programs may be seen as self-serving, and this may greatly impact potential and longevity of such programs."

American Fisheries Society Symposium 52:189–191, 2006

Opportunities for Linking Cooperative Research and Management: Partners' Needs and Interests

Gary L. Graham
Texas Sea Grant Program
Post Office Box 1125, West Columbia, Texas 77486, USA

What Will Managers Need from Cooperative Research and Management? What Will Fishermen Need? Scientists? Other Partners?

Although cooperative research in the southeast has been ongoing for many years, it has not been as significant or formal as in other regions. This is primarily due to funding allocations that have been directed to the Gulf of Mexico and South Atlantic fisheries. Cooperative research has primarily been conducted through Sea Grant programs, through the Gulf & South Atlantic Fisheries Foundation, and recently in NOAA Fisheries. If a valid commitment exists to expand formal cooperative research in the Gulf of Mexico and the South Atlantic, a need exists to educate the entire fishing industry, both commercial and recreational sectors, to its implications and structure. The same concept of educating the industries should be applied to cooperative management. Comanagement, as practiced in other regions and foreign countries (i.e., New Zealand), is not understood by the majority of industry members. Educational efforts should be applied to expand this concept if genuine interest exists from management and scientific entities.

Linkage among collaborators is an important part of initiating successful cooperative research and management. Once a project has been defined, it is vitally important that both parties consider the research project to be important and have a clear understanding of the expectations of each other. It is my opinion that not every scientist or every fisherman is suitable for cooperative research. It is important, however, that those who are be linked up. This important coupling can take place through those knowledgeable with the fishing industry and the scientific community—nongovernmental organizations (NGOs), universities, regulatory agencies, and so forth. Gatekeepers should be identified and utilized. An example of linking expertise is reflected with a recent Sea Grant research project in Texas. Professors in the Oceanography Department were awarded a project to determine impacts of trawling in the bay system relative to suspension of bottom sediments and pollutants. A portion of this study was to determine potential damage of shrimp trawls to bottom habitat. Needless to say, there was some hesitancy by industry to collaborate in this investigation. After I conferred with an industry NGO and discussed the importance of utilizing typical gear versus gear utilized by scientists not familiar with trawls, fishermen volunteered to provide vessels and equipment to collaborate in the study. A meeting to discuss the investigations was conducted, which included the scientists and collaborating fishermen. Habitat and seabed conditions were agreed upon in addition to experimental design. Conclusions from the investigations indicated that episodic events (i.e., storms and cold fronts) impacted suspension of seabed sediments in the bay more than effects of trawling. This is only one example of proper linkage between industry and the research community.

> "If a valid committment exists to expand formal cooperative research in the Gulf of Mexico and the South Atlantic, a need exists to educate the entire fishing industry, both commercial and recreational sectors, to its implications and structure."

In regards to cooperative research and management, it should be pointed out that industry collaborators should not be led to believe that cooperation will necessarily yield results that they want to see. It should be stressed that cooperative efforts yield results that industry and the scientific community have worked on together and that industry has input regarding the structure of scientific investigations being performed. Although they may not like the outcome of the investigations, the industry has an important role in seeing that the research is conducted in a manner in which they agree is acceptable. Cooperative research can provide a vehicle for positive relationships between the scientific/management community and recreational and commercial fishing industries. These bonds can assist in solving problems identified through cooperative studies and have proven to instill trust in the investigations.

Who Else Should Be Involved in Cooperative Research and Management Activities?

Until recently, only sparse participation in cooperative research has occurred in the recreational sector. This is not an indication for lack of interest from this expansive fisheries sector. Instead, it reflects a paucity of funding and necessary resources needed to conduct collaborative efforts.

The presence of the petroleum industry in the Gulf of Mexico has created several unique opportunities for cooperative research. Currently, a number of petroleum platforms exist that are scheduled to be abandoned. Opinions differ among various fishery user groups regarding the fate of these structures. Some groups believe that they should be partially disassembled and left in place as artificial reefs. Other groups want to see them removed completely from the gulf. Opportunities may exist to involve the petroleum industry in cooperative research for determining the benefits or disadvantages of utilizing these structures for artificial reefs. Other issues exist that provide potential cooperative efforts with the petroleum industry, such as impacts of marine organisms from water entrainment by open looped liquefied natural gas facilities. Construction of these facilities is becoming a common occurrence.

“Although they may not like the outcome of the investigations, the industry has an important role in seeing that the research is conducted in a manner in which they agree is acceptable.”

Involvement of environmental and conservation organizations in marine issues and policy has increased in recent years. These groups should be considered as players in cooperative research and management.

What Topics or Issues Are Ripe for More Cooperative Research or Cooperative Management?

The fishery for red snapper *Lutjanus campechanus* in the Gulf of Mexico reveals a unique example of a recondite fishery to manage. The stock is impacted from a diverse group of fishermen. A directed commercial fishery is highly restricted by catch allocations and size limits of this valuable species, but discard mortality resulting from this fishery is considered to be a major problem. A huge recreational fishery also targets this species, as well as recreational charter boats. The large numbers of the recreational component of this fishery is believed to impact rebuilding of the stock. A shrimp fleet of more than 2,700 vessels are licensed to fish federal waters, and it receives intensive indictment for harvesting juvenile red snapper that inhabit some of the shrimp grounds. Cooperative research grants have been directed toward solving the dilemma of overharvest, but it is believed that additional support could benefit the perplexity of managing this fishery. An opportunity to develop more intensive cooperative research and management strategies currently exists.

Presently, the number of potential cooperative research projects for the southeast region is robust. Clearly, we need more precise information on discard mortality of reef fish, specifically in the commercial red snapper directed fishery. Fishermen from this group have indicated for a number of years that they would collaborate in such a study. The potential for cooperative research that addresses management issues is extremely great in the recreational fisheries and charter boat communities. Although it has received support for addressing the bycatch issue, the shrimp fishery continues to struggle with the difficulties that it encounters with bycatch reduction. Excellent cooperators from the shrimp industry exist, which are dedicated to assuaging this dilemma. Another problem associated with the shrimp industry was identified several years ago and relates to a need to improve effort information. As a result of cooperative research, clear progress has been made to utilize new methods for collecting shrimp effort. A definite need exists to expand collection of effort data through cooperative research and examine possibilities for utilizing these data in cooperative management scenarios.

"...many scientists and managers find the culture of commercial fishermen difficult to understand, while fishermen sometimes are hesitant to work with the scientific/management community."

Does Cooperative Research and Management Require a Change in Governance Philosophies, or Can They Be Discrete Programs within the Current Government Structure? Are There Core Values Underlying Their Vision and Missions?

In regard to cooperative research, many scientists and managers find the culture of commercial fishermen difficult to understand, while fishermen sometimes are hesitant to work with the scientific/management community. Nonetheless, there exists within government excellent examples of personnel who have worked very successfully with the fishing industry. Concern exists that many in government consider collaborative research efforts to be second-class endeavors. There have been indications that those in government who participate in cooperative research may not advance in the governmental hierarchy as rapidly as those who perform esoteric investigations independently.

A firm decision needs to be established as to the degree of commitment that will be directed toward cooperative research and management. If this commitment has been established, then definitive guidelines need to be formulated that state the level of professionalism that is identified with those involved with cooperative research. It is unclear if this procedure has been performed.

It is my personal opinion that NOAA Fisheries has made tremendous strides in establishing a formal cooperative research program in recent years. I would prefer a formal structure regarding this program, but discrete support that has been provided in the southeast has been responsible for several highly visible and successful programs.

American Fisheries Society Symposium 52:193–196, 2006

Cooperative Management in the Great Lakes: Exploring Characteristics of Success

Charles C. Krueger
Great Lakes Fishery Commission
2100 Commonwealth Boulevard, Suite 100
Ann Arbor, Michigan 48105, USA

Roy A. Stein
Aquatic Ecology Laboratory
Department of Evolution, Ecology, and Organismal Biology
The Ohio State University
1314 Kinnear Road, Columbus, Ohio 43212, USA

Introduction

Our paper describes the unique nature of cooperative fishery research and management in the Great Lakes and formulates operational procedures based on our experiences that may be useful elsewhere.

Cooperative Great Lakes Fishery Management

The joint strategic plan

To coordinate fisheries management in the Great Lakes, the eight Great Lakes states, the province of Ontario, two tribal fisheries agencies, and the federal governments of Canada and the United States have adopted "A Joint Strategic Plan for Management of Great Lakes Fisheries" (joint strategic plan; GLFC 1997). First signed in 1981 and subsequently revised in 1989 and again in 1997, the joint strategic plan is the framework for fishery management decision making based on sound science and consensus. The goal statement of the joint strategic plan is

> To secure fish communities, based on foundations of stable self-sustaining stocks, supplemented by judicious plantings of hatchery-reared fish, and provide from these communities an optimum contribution of fish, fishing opportunities and associated benefits to meet needs identified by society for: wholesome food, recreation, cultural heritage, employment and income, and a healthy aquatic ecosystem.

Four strategies were adopted to achieve this goal: ecosystem management, consensus, accountability, and information sharing. The philosophy underlying the joint strategic plan is that each of the Great lakes fishery management agencies has a share in the Great Lakes ecosystem and that, in the interest of the common good, management authorities must be cooperative to be effective, rather than unilateral in their actions.

The strength of the joint strategic plan in its implementation comes through the activities of the lake committees. One committee was established for each Great Lake (Eshenroder 1987). A council of lake committees was also formulated to discuss basin-wide issues. Lake committees consist of one representative from each of the state, provincial, or tribal agencies with authority for fishery management. Federal signatories to the plan are not members but serve in an advisory role to the lake committees. Fundamentally, lake committees seek to manage each lake as a whole system, as opposed to management that follows the political fragmentation of the basin. Within the lake committees, decisions are made by consensus about the nature of desired fish community structure and the actions necessary to achieve them, stocking strategies, law enforcement activities, and harvest rates. Each management agency is then accountable for implementing the shared decisions within its own jurisdiction. The joint strategic plan is generally recognized as one of the most effective vehicles for cooperative, international fisheries management in the world.

Two unique aspects exist with the Great Lakes fishery management structure described above. First, agency involvement with the lake committee process is voluntary among the fishery management jurisdictions. No legislation forces or coerces the provincial, state, or tribal agencies to participate. No agency by its participation forfeits its jurisdiction or responsibilities. The second unique aspect of this voluntary, consensus-based approach is that international fishery management occurs among subnational governments (province, states, tribes) in the absence of central federal authority. In other international fisheries, federal governments usually have lead roles with subnational governments, assisting them in the process. In some cases, federal governments formulate agreements and establish international fisheries commissions with management authority to establish regulations and harvest quotas. In the Great Lakes, the United States and Canada established an international fisheries commission, but not with the same level of authority as in some other international fisheries commissions.

Great Lakes Fishery Commission

In 1955, the governments of the United States and Canada signed the "Convention on Great Lakes Fisheries," the treaty that established the Great Lakes Fishery Commission. This binational convention created the commission "to formulate and implement a comprehensive program for the purpose of eradicating or minimizing the sea lamprey populations" in the Great Lakes and to formulate, coordinate, and undertake a fishery research program that focused on the fish stocks of common concern (GLFC 1955). Thus, the responsibilities of the commission can be summarized first as the implementation of management program focused on control of the nonnative sea lamprey *Petromyzon marinus*. Beyond management authority for sea lamprey, the GLFC does not have any jurisdictional authority for fishery management in the Great Lakes. Second, the commission is to coordinate, conduct, and communicate fishery research. Third, the commission supports the activities of the agencies with jurisdictional authority to coordinate fishery management and to promote the productivity of Great Lakes fisheries. Given these responsibilities, the commission promotes both cooperative management and research. The commission was the logical organization to help facilitate the effective implementation of the joint strategic plan among its signatories and provide administrative support (e.g., schedules, communications, agenda setting) to the lake committees in their activities.

"The challenges facing Great Lakes fishery management are large within the context of the ecological and political diversity of the basin."

Challenges, conflicts, obstacles, and solutions

The challenges facing Great Lakes fishery management are large within the context of the ecological and political diversity of the basin. Invasions of foreign species from plankton to fish, allocation of percid harvest in Lake Erie, restoration of native species such as lake trout, lake sturgeon, and the American eel, and disagreements over the jurisdictional authority of some aboriginal groups are just a few of the challenges. Though we cannot comprehensively discuss all challenges, we can offer a few personal observations.

Goals and objectives are important!

Great Lakes fishery managers have worked hard over the past two decades to clearly describe fish communities and their associated fisheries, as well as articulate the characteristics desired in the future. These characteristics or management goals are known as fish community objectives and are developed through a consensus process by lake committees (e.g., Ryan et al. 2003). Species-specific plans also are developed for native species requiring specialized rehabilitation approaches, such as lake trout and lake sturgeon (Schneider et al. 1983; Krueger and Decker 1999). These species specific plans are considered to be nested within, and consistent with, the fish community objectives. Agreed-upon goals can resolve potentially negative interactions and serve to maintain a laser focus on what management is attempting to achieve. Conflicts in day-to-day management are reduced when lake managers across jurisdictions share a common understanding of their long-term goals. Conflicts do arise when discussions focus on management actions such as harvest regulation and are not explicitly linked to agreed-upon goals. This situation happens when the participants do not personally agree with established goals or (have forgotten the earlier agreements made by their agencies). These situations can be summarized as "arguing about actions while being at cross-purposes on goals."

Knowing jurisdictional boundaries is important!

Cooperative management requires that the participants know their roles and how each fishery management agency is similar to, or different from, their own. Some fishery management agencies are organized within a natural resources agency that includes environmental protection, whereas others may be solely focused on fish and wildlife management. Participants in cooperating agencies must know not only bounds of their agency, but also those of other agencies. Jurisdictional authority is jealously guarded by individual agencies in the Great Lakes. For example, the staff serving the Great Lakes Fishery Commission may provide advice or offer rational dialog regarding ongoing fishery management; however, great care is given to only provide input and not infer authoritative direction. Should the commission begin to operate in a directive manner, the province, states, and tribes will ensure that the folly is identified and corrected immediately!

Personalities are important!

Many governmental agencies seem to relish internal reorganization. Reorganization may refocus the agency appropriately, but often, much energy is expended with a commensurate decline in morale. Individual people are a real key to effective cooperative management. Good people can overcome weak organizational structures. Great organization structures can be sabotaged by complacent, unmotivated people. Cooperative fishery management requires managers who see their role as resource stewards; they are not threatened by new information or approaches, enjoy discussion and debate, and care about fish and people. The best managers are those who lead based on guiding principles (e.g., honesty is the best policy, do no harm to the resource): have a stewardship concern for future generations' access to resources, have strong listening skills, recognize the importance of relationship building with managers outside their agency, and know that understanding another manager's position does not mean you agree with it. When at least one such manager exists in each of several agencies on a Great Lake, then they can become a leading force for progressive fishery management. Most outstanding leaders have a special charismatic ability to effectively communicate and motivate others. Loss of key leading fishery managers can be devastating. Deliberate succession planning for these managers must occur if long-term, successful, cooperative fishery management is to continue.

> "...the staff serving the Great Lakes Fishery Commission may provide advice or offer rational dialog regarding ongoing fishery management; however, great care is given to only provide input and not authoritative direction."

Communication is important!

Honest communication among managers is required for cooperative Great Lakes fishery management. For example, when determining harvest allocation among jurisdictions, agreements are made within the lake committee. Managers then report to their agencies. Sometimes the agency balks at this nonbinding agreement; similarly, fishery stakeholders may stage a rebellion against the agency. The manager's best course of action, under these circumstances, is to return to the lake committee forum and state the circumstances. Finding fault with the science supporting the earlier agreement guarantees a bad outcome. Honest communication builds predictability, trust, and mutual respect in relationships in cooperative management regimes.

Applying the Great Lakes Experience

The Great Lakes have an organizational structure where all management agencies have separate but equal voices, where stakeholders are advisory, where decisions are made by consensus, and where an oversight body with credibility (but no management authority) serves to arbitrate difficult issues. Because the system is voluntary, protects and respects individual jurisdictional authority, and requires written goals and objectives, it works. Finally, it works because strong, though cooperative, charismatic leadership characterize the managers within the basin. Their desire for consistent ecosystem management overrides local, more restricted perspectives, which are, at times, embraced by their administrators.

"Because the system is voluntary, protects and respects individual jurisdictional authority, and requires written goals and objectives, it works."

Cooperative research, in which agencies and universities forge partnerships, and in which funding agencies join forces to attack basin-wide problems, also is driven by effective organizational structures but most importantly by flexible, creative administrators. Within the Great Lakes, we have been indeed fortunate to have innovative administrators that serve to deal effectively with ever-evolving ecosystem issues. Our view of cooperative management and research and its attendant difficulties lies, first and foremost, with people ... success requires fishery managers that have courage and credibility ... critical character traits that enable managers to put fish first.

References

Eshenroder R. L. 1987. Socioeconomic aspects of lake trout rehabilitation in the Great Lakes. Transactions of the American Fisheries Society 116:309–313.

GLFC (Great Lakes Fishery Commission). 1955. Convention on Great Lakes Fisheries. Great Lakes Fishery Commission, Ann Arbor, Michigan.

GLFC (Great Lakes Fishery Commission). 1997. A joint strategic plan for management of Great Lakes fisheries. Great Lakes Fishery Commission, Ann Arbor, Michigan.

Krueger, C. C., and D. J. Decker. 1999. The process of fisheries management. Pages 31–59 in C. C. Kohler and W. A. Hubert, editors. Inland fisheries management in North America, 2nd edition. American Fisheries Society, Bethesda, Maryland.

Ryan, P. A., R. Knight, R. MacGregor, G. Towns, R. Hoopes, and W. Culligan. 2003. Fish-community goals and objectives for Lake Erie. Great Lakes Fishery Commission, Special Publication 03-02, Ann Arbor, Michigan.

Schneider, C. P., D. P. Kolenosky, and D. B. Goldthwaite. 1983. A joint plan for the rehabilitation of lake trout in Lake Ontario. Lake Ontario Committee, Great Lakes Fishery Commission, Lake Trout Subcommitteee, Ann Arbor, Michigan.

American Fisheries Society Symposium 52:197–203, 2006

Meeting Hawaii Longline Industry Needs to Reduce Fisheries Bycatch through Cooperative Research

SEAN C. MARTIN
Hawaii Longline Association
45-519 Mokulele Drive, Kaneohe, Hawaii 96744, USA

ERIC GILMAN
Blue Ocean Institute
2718 Napuaa Place, Honolulu, Hawaii 96822, USA

PAUL DALZELL
Western Pacific Regional Fishery Management Council
1164 Bishop Street, Suite 1405, Honolulu, Hawaii 96813, USA

Introduction

Hawaii pelagic longline fisheries are faced with strong economic and social incentives to reduce bycatch of sensitive species, including sea turtles and albatrosses. Cooperative research has proven an extremely successful approach to reduce seabird bycatch in Hawaii pelagic longline fisheries. Here, we describe the Hawaii longline industry's experiences with cooperative research, as well as alternative approaches to address fisheries bycatch, including litigation, commercial demonstrations, educational materials, industry capacity-building, institution of a pilot fleet communication program, and contribution to international conferences. We describe information of interest to industry from cooperative research and recommend that U.S. fishery management authorities use the Hawaii cooperative research experience as a model for broad nationwide replication.

The Hawaii Longline Fleet and Prominent Bycatch Issues

In 2004, there were 125 active Hawaii longline tuna and swordfish vessels, which made 1,338 trips, setting about 32 million hooks. Table 1 summarizes target species catch per unit of effort for the combined Hawaii longline tuna and swordfish fisheries from 1999 to 2004. In 2004, the Hawaii longline fisheries landed 18.5 million pounds and generated exvessel revenues estimated at $42.6 million with tuna *Thunnus* spp. the dominant components of longline landings.

TABLE 1. Hawaii pelagic longline tuna and swordfish fisheries catch per unit of effort (CPUE), number of fish per 1,000 hooks, 1999–2004 (U.S. National Marine Fisheries Service Pacific Islands Regional Office, unpublished data).

Year	Tuna CPUE	Sharks CPUE	Billfish CPUE	Other CPUE[a]
1999	9.21	4.59	3.9	4.8
2000	8.18	3.91	2.88	4.8
2001	8.64	2.1	1.61	4.21
2002	7.48	1.87	0.98	4.27
2003	6.33	2.32	1.77	4.58
2004	6.42	2.34	1.24	5.49

[a] Dolphinfish (also known as mahimahi) *Coryphaena hippurus*, opah (also known as moonfish) *Lampris guttatus* (also known as *L. regius*), oilfish *Ruvettus pretiosus*, sickle pomfret *Taractichthys steindachneri*, and wahoo *Acanthocybium solandri*.

The shallow-setting Hawaii longline swordfish fishery results in much higher levels of seabird and sea turtle bycatch than the deep setting longline tuna fleet. The Hawaii longline swordfish fishery was closed for more than 4 years and is now subject to strict management measures, including prescribed use of large circle hooks and fish bait, restricted annual effort, caps on turtle captures, 100% onboard observer coverage, required possession and use of specialized dehooking equipment, and mandatory attendance of an annual protected species workshop by vessel operators and owners, as a result of concerns over turtle interactions. If seasonal limits on turtle interactions are reached, the fishery is closed for the remainder of the calendar year, and if a threshold is exceeded, federal resource management agencies consult to determine if additional restrictions on the fishery are warranted. Furthermore, the Hawaii longline swordfish and tuna fleets are each authorized to annually take, through injury or mortality, one federally listed endangered short-tailed albatross. If more than one short-tailed albatross is observed to interact with gear of the Hawaii longline tuna or swordfish fleet in a single year, resource management agencies consult to determine if the fleet should be required to employ additional seabird avoidance measures. Fewer than 100 Laysan and black-footed albatrosses are now annually captured by the fleet. This is down from thousands of albatrosses being captured each year, before the fleet was required to employ seabird avoidance methods and restrictions on swordfish fishing effort were in place. The fleet has not had any observed captures of a short-tailed albatross. Since June 2001, management authorities have required the Hawaii longline tuna and swordfish fisheries to employ a number of measures intended to reduce seabird bycatch, including weighted branch lines, thawed and dyed bait, offal discards, and night setting in certain geographical areas for certain components of the Hawaii longline fleet. Interactions between the Hawaii longline fleet and false killer whales is another issue that has received recent attention. While there have been claims that this is causing population-level effects, in reality, there is very little understanding of the status and trends of false killer whale populations or the consequence of interactions with longline gear.

“Longline fishermen have a large repository of knowledge and information related to bycatch, which can be tapped to contribute to finding effective and practical solutions.”

Collaborative Research and Commercial Demonstrations to Reduce Bycatch

Between 1999 and 2003, the Hawaii Longline Association collaborated with fishery management authorities and an environmental organization to conduct three research experiments and commercial demonstrations of various strategies (blue-dyed bait, towed buoy, offal discards, streamer line, underwater setting chute, and side setting) to reduce seabird bycatch in longline gear. We have determined that several seabird bycatch avoidance methods are capable of nearly eliminating bird captures in pelagic longline fisheries when effectively employed. Our industry-led experiments have focused on identifying the most effective seabird bycatch reduction methods that are also economically viable and practical. Fishery management authorities are now amending regulations on measures for the Hawaii longline fleet to reduce seabird bycatch based on results from the most recent research.

Longline fishers are some of the most qualified people to develop and improve seabird bycatch mitigation techniques. Longline fishermen have a large repository of knowledge and information related to bycatch, which can be tapped to contribute to finding effective and practical solutions. This has been demonstrated by the successful collaborative research initiatives in Hawaii and elsewhere. Mitigation methods that effectively avoid seabirds do not reduce fishing efficiency, or, better yet, increase fishing efficiency and provide operational benefits, have the highest chance of being accepted by industry. The longline association became an active participant to address seabird bycatch problems by instituting and participating in research and commercial demonstrations and supporting adoption of regulations based on best available science before restrictions, embargos, and possible closures were imposed on the fleet. This bottom-up approach fostered a sense of industry ownership for effective seabird mitigation methods and resulted in high compliance with resulting rules requiring use of seabird avoidance methods. By being directly involved in the development and testing of seabird avoidance methods, Hawaii longline fishers developed a sense of ownership for these tools and now support their required use.

Litigation

Over the past 5 years, there have been numerous lawsuits filed against the U.S. National Marine Fisheries Service by environmental nongovernmental organizations and the Hawaii Longline Association generally over the bycatch of sea turtles, seabirds, and whales by the Hawaii longline fisheries. There have been a number of positive results from the litigation, but overall, this has not been a wise long-term approach or efficient use of money, time, or energy to address fisheries bycatch.

There was little attention paid to effects to reduce bycatch of sea turtles in the Hawaii longline fisheries until litigation aiming to close the fishery began in 1999. The threat this litigation posed to the fishery caused the fishing industry, working with regulators and managers, to develop methods that resulted in substantial improvements involving changes in fishing gear, fishing practices, and methods to handle and release caught turtles. These changes reduced sea turtle interactions and injuries. Turtle bycatch levels are now much lower than in the past, and turtles are being released with less injury and higher chance of survival.

Another positive result of the litigation was increased cohesiveness of Hawaii Longline Association members. The three main ethnic groups comprising the fishery came together to counter efforts to eliminate their source of livelihood and denigrate the reputation of the Hawaii longline fisheries. The industry is now in a much better position to represent their interests.

However, even after substantial improvements were adopted by fishery management authorities and the longline industry, the litigation continued, as some environmental groups continued to pursue their goal of permanently closing the fishery. The result was that the fishers became bitter, were much less receptive to collaborating with outside groups, and lost the drive to pursue voluntary initiatives to innovate new bycatch solutions, which might also be exportable to longline fleets internationally. Other environmental groups with a goal of reducing fisheries bycatch and reducing a source of turtle mortality by working with fishers had a much more difficult time gaining industry's trust to work with them as a result of the actions of the groups that were working to close the fishery. Groups that wanted to pursue collaborative work with the Hawaii longline fleet to make the Hawaii fleet a model fishery, and to export identified solutions internationally, were frustrated by the misplaced efforts to close the Hawaii fisheries.

“The three main ethnic groups comprising the fishery came together to counter efforts to eliminate their source of livelihood and denigrate the reputation of the Hawaii longline fisheries.”

The efforts to close the Hawaii fleet may have actually increased turtle and bird mortality. During a 4-year closure of the Hawaii longline swordfish fishery due to concerns over bycatch of sea turtles, swordfish supply to the U.S. marketplace traditionally met by the Hawaii fleet was replaced by imports from foreign longline fleets, including from Mexico, Panama, Costa Rica, and South Africa, which have substantially higher ratios of sea turtle captures to unit weight of swordfish catch and less stringent or no measures to manage seabird bycatch.

The Hawaii Longline Association spent more than US$1.6 million and enumerable staff hours over the past 5 years as a result of involvement in this litigation. If this money, plus the funds spent by the U.S. government and environmental groups on the litigation, had instead been used to conduct research to find effective and commercially viable solutions in the Hawaii fleet and abroad, this might have saved many more turtles' lives.

In the Hawaii longline fisheries, collaborative, industry-led research has resulted in substantially more progress towards identifying effective and practical solutions to seabird bycatch than litigation has at identifying solutions to turtle bycatch. Furthermore, the collaborative research cost an order of magnitude less than that spent on lawsuits.

Economic Viability, Practicality, and Enforceability Considerations in Designs of Cooperative Research

The research experiments on techniques to reduce seabird bycatch in the Hawaii longline fisheries provide an example of how research can be designed to collect information on economic viability, practicality, and enforceability. Analyzing differences in alternative seabird avoidance methods' effect on bait retention, hook setting rates, and target fish catch per unit of effort; operational benefits and costs; time and money to adopt and employ; and enforceability is of high interest to industry, fishery management authorities, and other stakeholders.

Given the political context and management frameworks of the majority of the worlds' longline fisheries, there is a need to focus on the commercial viability of bycatch reduction methods in order to catalyze changes in fishing methods and gear and regulatory measures that will reduce longline bycatch. To resolve global fisheries bycatch problems, there is a need to identify and institute the broad use of methods that not only have the capacity to minimize bycatch of sensitive species, but that are also practical and convenient and provide crew with incentives to employ them consistently and effectively. It is critical to account for economic and social values of longline fisheries to achieve changes that reduce bycatch.

For instance, as the loss of bait to seabirds and concomitant reduction in catch of fish can be significant, the use of seabird avoidance measures is expected to be cost saving for longline fisheries. However, most longline fleets do not employ effective seabird avoidance methods, despite the availability of effective methods that also increase fishing efficiency (Brothers et al. 1999a; Gilman 2001; FAO 2003). Reasons for this may be (1) low industry awareness of the availability, effectiveness, and practicability of these seabird avoidance methods; (2) few national fishery management authorities manage interactions between seabirds and longline vessels or require employment of effective seabird avoidance methods (Brothers et al. 1999a; BirdLife International 2003; FAO 2003; Gilman and Freifeld 2003); and (3) lack of a sufficiently strong economic incentive for industry to change long-standing fishing practices. Recognizing that this context of global longline fisheries applies to many global commercial marine fisheries, maximizing industry's sense of ownership for using effective bycatch avoidance measures and providing industry with incentives for voluntary compliance are needed. Commercial fishing industries responds best to economic incentives and disincentives (Gilman et al. 2002). Bycatch mitigation methods that increase fishing efficiency and have operational benefits have the best chance of being accepted by industry. Ecolabeling and certification programs can also provide industry with strong market-based and social incentives to meet criteria to be certified as a sustainable fishery, including the employment of effective bycatch reduction methods, but requires adequate marketing of the label to make it economically viable for industry to participate (Gilman et al. 2002). Additionally, if regulations requiring the use of bycatch avoidance methods are effectively enforced and carry sufficient economic consequences for noncompliance, broad industry compliance can be achieved.

“...the collaborative research cost an order of magnitude less than that spent on lawsuits.”

Outreach, Capacity-Building, and Disseminating Lessons Learned

The Hawaii longline association, in partnership with fishery management authorities and environmental conservation groups, has produced a number of educational materials on methods to reduce fisheries bycatch. These include a poster and pamphlet on side setting to reduce seabird bycatch, a poster on best practices to handle and release incidentally caught seabirds in longline gear and methods to reduce seabird capture, and a booklet on methods to reduce sea turtle bycatch in pelagic longline gear and practices to handle and release captured turtles. We produced these educational materials in several languages to enable the various ethnic groups in our fleet, and longline fisheries abroad, to access the information. The Hawaii Longline Association is also able to disseminate lessons learned from research experiments and commercial demonstrations and learn from other fisheries bycatch research through participation in and providing financial support for conferences such as the International Fishers Forum series.

> "We produced...educational materials in several languages to enable the various ethnic groups in our fleet...to access the information."

The Hawaii Longline Association is working with management authorities and an environmental organization to initiate a dockside training program for longline vessels to convert deck designs from conventional setting position from the stern to the side of the vessel to reduce seabird bycatch. Deck conversion will require considering the deck position for setting; selection of main line shooter hinges and hydraulics, line pullers, motor and mounting plate design for starboard setting; and design, construction, and installation of a bird curtain. Technical assistance will also be available to captains and crew on best fishing practices for setting from the new position, including timing for clipping branch lines to the main line and practices for baited hook throwing.

Because most bycatch species are highly migratory (frequently move in and between national jurisdictions and interact with longline vessels on the high seas), international collaboration is necessary to effectively address fisheries bycatch problems. As a global and multinational problem, the solution to seabird and sea turtle mortality in longline fisheries will require international collaboration by management authorities, industry, and other stakeholders to share technical and financial resources to augment capacity. These education and outreach programs are an investment to bring about changes in behavior and attitudes by having a better informed industry of prescribed fisheries bycatch avoidance methods and, in some cases, operational benefits from employing these techniques. Showcasing the results of industry-led research to reduce fisheries bycatch also has the benefit of broadly disseminating the results so that the effective methodology can be replicated in other fleets worldwide and ineffective components can be improved.

One Fleet Pilot Project

The Hawaii Longline Association worked with an environmental organization and fishery management authorities to research the state of knowledge of employing fleet communication programs to reduce fisheries bycatch and is now planning to institute a pilot program to reduce bycatch of sea turtles and albatrosses. Instituting a fleet communication system to report near real-time observations of bycatch hotspots enables a commercial marine fishery to operate as a coordinated "One Fleet" to substantially reduce fleet-wide capture of protected bycatch species, including fish, seabirds, sea turtles, and marine mammals. This benefits the bycatch species per se, reduces waste, and can provide economic benefits to industry by reducing the risk of exceeding government-established seasonal bycatch thresholds, as well as avoiding possible future declines in target species catch resulting from bycatch of juvenile and undersized individuals. We analyzed case studies of fleet communication programs of the U.S. North Atlantic longline swordfish fishery, U.S. North Pacific and Alaska trawl fisheries, and U.S. Alaska demersal longline fisheries. Available information from these case studies supports the inference that they have substantially reduced fisheries bycatch and provided large economic benefits that outweigh relatively nominal operational costs.

It is not yet known how likely it is that the Hawaii longline swordfish and tuna fleets will annually exceed seasonal sea turtle bycatch limits, making it difficult to assess if economic benefits from instituting a "One Fleet" protocol, resulting from enabling the fleet to operate for a longer time period, will outweigh economic costs from managing the fleet communication program. Furthermore, it might not be possible to definitively determine the effect of instituting the fleet communication program on sea turtle and seabird bycatch rates because there might not be a suitable control for comparison. Historical bycatch rates would not provide a suitable comparison because the fleet is now using different technological methods designed to minimize seabird and sea turtle bycatch. Furthermore, comparison of bycatch rates from different time periods can be confounded by numerous variables, including weather, seabird and turtle behavior, fishing practices, location of fishing grounds, and consistency in observer methods. However, if some of the Hawaii longline vessels opted not to participate in the fleet communication program, a comparison of bycatch rates of participating and nonparticipating vessels could provide an understanding of the effect on bycatch rates from this single factor, assuming that there are no other substantial differences between the two categories of vessels, as was possible for the Alaska demersal longline fisheries fleet communication program. Nonmonetary benefits to the Hawaii longline industry from instituting a "One Fleet" program to reduce turtle and bird bycatch could be substantial, such as from positive media coverage and other values not described by established monetary indicators.

> "...fishers are some of the most qualified people to develop and improve bycatch avoidance strategies."

Conclusions

In Hawaii and elsewhere, we have seen that fishers are some of the most qualified people to develop and improve bycatch avoidance strategies. Fishers have a large repository of knowledge and information related to bycatch, which can be tapped to contribute to finding effective and practical solutions. Mitigation methods that effectively avoid bycatch, do not reduce fishing efficiency, or, better yet, increase fishing efficiency and provide operational benefits have the highest chance of being accepted by industry. Fishers and fishery associations need to become active participants to address bycatch problems by participating in research and commercial demonstrations, implementing best practices, and supporting adoption of regulations based on best available science before restrictions, embargos, and possible closures are imposed on them.

Most countries have a low degree of political will to address fisheries bycatch problems and, as is the case in Hawaii, have scarce resources for enforcement of bycatch management measures. Few national fishery management authorities have frameworks to manage interactions between sensitive bycatch species and fishing vessels and do not require employment of effective bycatch avoidance methods. A bottom-up approach that fosters a sense of industry ownership for effective bycatch mitigation methods, and concomitant compliance with requirements for using bycatch avoidance methods, is needed in these countries. Fishers are among the most qualified people to innovate mitigation methods and should be encouraged to develop and test bycatch avoidance methods. In this way, industry develops a sense of ownership for these tools and supports their required use. Collaboration by fishing industry, government, and environmental organizations is the most effective approach to reduce fisheries bycatch and can be accomplished with strong and enlightened leadership and willingness to put differences aside to work towards a common good.

While the effectiveness of this approach to address fisheries bycatch is broadly recognized, there has been much too small an amount of funding allocated for cooperative research and commercial demonstrations to find solutions to sea turtle, seabird, and other bycatch problems in longline gear. In the United States, this is a result of the government's fear of being sued if they propose to conduct or fund experiments in U.S. fisheries that result in injury to protected resources, even though these experiments will potentially result in substantial reductions in mortality of these species when best practices are identified and spread to multiple fisheries. Some U.S. fishery management authorities are funding experiments to test technical measures to reduce sea turtle bycatch in longline fisheries abroad, in part to avoid problems with trying to receive permits and risk being sued by conducting the research in domestic fisheries, but too little research is being supported, there is no coordination and efforts are duplicative, and solutions found abroad may not be relevant to domestic fisheries. The amount of research being conducted is too small, research needs to occur at home to find solutions that we can have confidence will work in our fisheries, and the agencies designing the experiments need to do more to tap into fishers' knowledge to identify new promising strategies.

"...the agencies designing the experiments need to do more to tap into fisher's knowledge to identify new promising strategies."

American Fisheries Society Symposium 52:205–209, 2006

Session 3B: Opportunities

Linking Cooperative Research and Management: Integration, Panel Summary

JIM D. MURRAY
Sea Grant Extension , National Sea Grant College Program
NOAA/OAR, R/SG, Room 11752, SSMC3
1315 East-West Highway , Silver Spring, Maryland 20910, USA

Feedback needs to go to the managers and to the researchers so that we can redirect research or change the research that we are working on… Cooperative research has a place in every feedback loop. Should eventually be able to drop the word cooperative and it will become implicit in the world of research because it has been fully integrated. [Bonnie Ponwith, acting director, Office of Science and Technology, NOAA Fisheries]

The final session of the joint Sea Grant/American Fisheries Society symposium focused on the integration of cooperative fisheries research and management. Jim Murray, director of outreach with the National Sea Grant Office, moderated the panel and noted that a lot of progress has been made in the past 20 years in the fields of cooperative research and management. The councils are examples of cooperative management, but 20 years ago, the council process did not work very well. In the past, the participation process was largely ritualistic—the council developed fisheries management plans (FMPs), there was a law requiring public hearings, fishermen came to public meetings and they would angrily vent, council members would speed the meeting ahead, and the council would move forward with very little change to the whole FMP as a result of the whole process. This process has changed. The take-home point is that change is often incremental in society and government, but progress is being made.

Below are some speaker highlights, followed by a summary of the facilitated panel-audience discussion. White papers prepared by panelists follow this panel summary.

Peter Kendall, FV *Miss Alicia*, Portsmouth, New Hampshire

- It would be nice if scientists and fishermen got together before a meeting to find a common goal and theme and decide ahead of time. It would achieve a lot more if you can work together on an agreed common goal before going through the process.
- We can only gain from cooperative research and management. There are hurdles that need to be addressed though, including a continued disconnect between funding agencies, fishermen, industry, scientists, and management.
- Need to work toward taking a project's

results and translating them into practical guidelines and using the ideas to formulate a management response. In order to do this, we need liaisons between funding agencies and management agencies, someone who can take project results and put them into management. Research projects should be picked and chosen by the management agencies so that research projects are scripted in such a way that they can use the results in the decision-making process in a timely matter.

- A hurdle for scientists *and* fishermen—funding problems/issues. Scientists are always trying to get money (very competitive), and thus there is not as much shared information as there should be.

Integration is a Process

For over the past 20 years, the integration of cooperative research and comanagement systems in U.S fisheries management has progressed, albeit slowly. Yet, the potential for further integration

- Effective fishery management requires (a) a foundation of credible scientific information, (b) a consistent flow of high-quality information, (c) managers that understand the science well enough to use it in making management decisions, and (d) active engagement of relevant sectors of society.
- Research feeds a product to managers (science information and advice); the managers use this information and produce an outcome (decision). Managers are the clients of science—the management process is the driver for determining what science needs to be done.
- Science must be responsive and timely to meet management needs, and a feedback loop must be established to enable the science programs to adapt to changes in management.

Bonnie Ponwith, acting director, Office of Science and Technology, NOAA Fisheries (substituting for Steve Murawski, director of scientific programs and chief science advisor, NOAA Fisheries)

- Setting research priorities is both an ecological and social phenomenon. Competition is an important tool for setting research priorities.
- Cooperative research should be fully integrated into comprehensive research planning and execution.
- Science products must meet rigorous peer-review standards.
- Cooperative partners play an important role in facilitating communication and helping to bridge gaps in communication.
- Integrating research and management must take place at the onset in the priority setting stage, not after the research is completed.

Robert Pomeroy, associate professor and Sea Grant Fisheries Extension specialist, University of Connecticut, Avery Point Campus

- Comanagement is a paradigm shift in governance and management. It is a society-centered approach of self-governance, participation, networking, and decentralization.
- Comanagement is a partnership agreement between fishermen, government, and the stakeholders, involving shared authority of the fishery, consultations and negotiations, and clearly defined roles and rights among partners.
- Successful power sharing takes those that are not empowered, usually the fishers, and allows them to work on an equal basis to those that are in power.
- A lot of it is dealing with conflict, so you need a process or forum to make rules, solve problems, learn, and open up lines of communication to help deal with conflicts and a process to build institutions that work more effectively.
- Cooperative research and comanagement are different but must work together, compliment and support each other, inform management, nourish community education, and foster community involvement.
- Empowered people become more involved in the management process itself and increase the chances that comanagement programs will succeed through local involvement and ownership. Scientific and local knowledge can compliment and enhance each other.
- Empower community members to utilize local knowledge and practices and give the local community the opportunity to learn skills and learn about the research process.

remains tremendous. The panel made clear that there is no single blueprint or model for integrating cooperative research and comanagement systems, but rather, these systems must evolve, adapt, and adjust over time. Integration is a process that builds on trust, credibility, and previous successes, involving the integration of research and management priorities, scientific and local knowledge, funding, and coordinated institutions. Integration is specific to political, managerial, and scientific contexts.

Characteristics of Successful Integration

The panel identified some common characteristics of successful integration systems. A central tenet of any cooperative research and comanagement scheme is that the scientific product must be of the highest quality and meet peer-review standards. To maximize success, it is essential that all relevant stakeholders work together from the outset to identify goals, objectives, and approaches for addressing mutually acknowledged fishery research and management needs. Successful integration not only requires a mutual understanding and agreement of the desired outcomes, but perhaps more importantly, a degree of trust among participants that provide integrity in the process. Third-party inter-

Jason Schratwieser, conservation director, International Game Fish Association

- The International Game Fish Association has been around for a while, since 1939, and was created for two purposes: (1) to create a code of ethical angling rules that allow for comparison of catches worldwide, and (2) to promote and encourage the sport of fishing for both recreation and research, as a source of data.
- Problem: cooperative management is difficult because of all the different user groups with different backgrounds, which often leads to distrust.
- Solution: get appropriate groups involved earlier in the research and data collection—this will foster trust and transparency and this allows for trust from the beginning.
- There is a need for better cooperative management, implementation, and compliance.

mediaries (university-based organizations, nongovernmental organizations, professional project coordinators, facilitators or mediators, and Sea Grant extension agents) can play an important role in convening key stakeholders and building support and trust. These third parties could also fill an important liaison function to link research results to practical management applications. Establishing clear priorities that guide research funding decisions is also essential. A clearly articulated research plan that is directed at critical management needs can serve to broaden the funding base by providing a common frame of reference for funding agencies.

Future Focus

As cooperative fisheries research and management evolve, exciting new opportunities will arise. Historically, comanagement has worked best in fisheries that can be compartmentalized at a finite or small scale. As fisheries management moves towards an ecosystem-based management system, how will comanagement systems enable a holistic regional approach to fisheries management? What institutional and legal changes will be needed to allow for fisheries management systems to adapt to emerging comanagement opportunities? How can fisheries managers and researchers utilize and integrate emerging advances in ocean observing systems to support cooperative research and comanagement programs? Most importantly, how can we design performance and evaluation protocols to ascertain whether or not we are achieving intended outcomes by using the most cost-efficient methods?

As we look to the future, the public resources available to support fisheries research and management at the federal, regional, and state levels will continue to be constrained. By implication, effective management of fisheries resources mandates that affected parties must embrace the integration of

Steve Waste, program manager for analysis and evaluation, Fish and Wildlife Program, Northwest Power and Conservation Council

- Difficulty in agreeing on problem definition coupled with shared funding responsibilities under overlapping mandates have resulted in a fragmentation of research efforts in the Columbia River basin.
- Key research questions have persisted for years because resource management agencies have been unable to secure or collaborate on the funding commitments necessary to mount organized, large-scale, and/or long-term field experiments to address them.
- The Columbia River Basin Research Plan recommends that policy makers facilitate the integration of the currently compartmentalized research agendas and budgets of entities that share common objectives, by convening a regional research partnership.
- The partnership would provide a forum for the identification of shared research priorities and development of collaborative implementation and funding strategies.
- Developing standardized approaches, making and securing long-term funding commitments, and coordinating with regional partners are all essential to the development of an integrated, regional approach to research.

cooperative research and management. As suggested by panelists, using third-party liaisons, including outreach and education as part of the funding criteria, and designing cooperative research plans that have direct and focused management implications will contribute to the success of cooperative programs.

American Fisheries Society Symposium 52:211–212, 2006

Integrating Cooperative Research and Management

Peter Kendall
FV Miss Alicia
159 West Road, Rye, New Hampshire 03870, USA

My opinions in this paper come from having been in the fishing industry for 20 years. Currently, I am the owner and operator of two inshore boats. I have participated as the co-principle investigator on a number of cooperative research projects. Previous to owning two boats, I was the manager of a fishing cooperative. I have dealt with fishermen, management, and programs that seek to bridge the gap between them using cooperative research. Throughout my 20 years in the industry, I have always played an active role in management decisions. I have sat on advisory panels and also expressed my views in public forums. I have also participated in the grant-funding process by reviewing grant proposals. With all these experiences, I take an open-minded approach to cooperative research and cooperative management because we can only gain from them.

How Do Cooperative Research and Management Inform and Guide Each Other? What Opportunities Exist for the Future?

At the present time in New England, there is still a *disconnection between funding agencies, fishermen, scientists, and management.* Why? All identities do not know how to take the results of a research project and translate it into a practical guideline or strategy to formulate a management response. The science is sound, but applying it towards these issues can be difficult. A number of projects *have* been used as a tool or strategy; however, the process takes a long time (too long). Before a project is funded the question should be asked—*how easily and practical the results can be derived and how quickly can they be applied in shaping new regulations.*

The funding agencies and management should each have a *liaison* that can give each other guidance as well as feedback regarding research projects. With all the money being spent on cooperative research, the two groups should coexist (for they share a common goal). Management should be able to direct a certain theme or topic that funding agencies award for research projects. This will target those areas in the fisheries that are the most important or time sensitive. If the core of the research projects is what the management is focusing on at that time, the more apt the management is in using the results of a project in its decision-making process. This liaison should be responsible for reporting the project, the goal, and the results on each research project and, together with management, determine if it could be applied in the decision-making process.

“The funding agencies and management should each have a *liaison* that can give each other guidance as well as feedback regarding research projects.”

I feel that, although all funded projects go through a process to get approved, some results may not be implemented because management does not know how to apply the science that was derived from it. Projects that have shown sound results (positive or negative) should be used in a more timely fashion to help shape management decisions.

There is also a *disconnection between scientists.* In this highly competitive field, it is hard for scientists to share ideas and information. Once the results of a project are published, it is hard to apply them toward management decisions. In the future, we need to work more closely together to achieve a common goal.

What Are Future Research Opportunities, Needs?

Opportunities

Research opportunities should be shaped around what future issues all the stakeholders deem important to that particular region or sector.

Needs

1. Research needs in New England are starting to shift toward an ecosystem-based approach. It does not matter if it is fish or phytoplankton; there is strong support to find the answers to how one species has an effect on another.
2. Habitat is another high priority topic of discussion. As habitat is the base of an ecosystem, how it is affected by change is of great interest.

How to Get Cooperative Research or Cooperative Management Started in a New Place?

A lot can be learned from existing programs around the country and globe. I do not think there is a perfect model or format to follow because most management decisions are region or species sensitive. One can take positive and negative input from existing programs and apply them as needed in a specific region.

Some main points to consider when first starting cooperative management and/or research programs are as follows:

1. *Have a common goal between all stakeholders.* This is the hardest step in shaping cooperative management. To achieve this goal, there need to be tools available to be flexible. As shown in the past, cooperative management needs to be able to adjust as new science and data become available.
2. *Bring all stakeholders to the table (no one should feel alienated from the process).* Conduct meetings so that people can communicate with each other in an open format. It is my opinion that the more personal the communication is, the more positive the results are. Publications, management meetings and court systems are not the places to be communicating with one another. This is counterproductive to the common goal we are all striving for.

In conclusion integrating cooperative research and management can be done. It is going to take time and effort by all parties. The path will be bumpy at times, but in the end, we will all benefit.

American Fisheries Society Symposium 52:213–214, 2006

Ensuring the Scientific Integrity of Cooperative Research

Bonnie J. Ponwith
NOAA Fisheries Service, Office of Science and Technology
1315 East West Highway, SSMC3, 12th Floor, Room 12451
Silver Spring, Maryland 20910, USA

There is a growing demand for timely, high-quality science for use in managing the nation's living marine resources. Leveraging limited research funds through research carried out cooperatively is a means of meeting that demand. Research conducted by agency scientists in cooperation with the commercial and recreational fishing industry, nongovernmental organizations, and universities contributes significantly to the overall knowledge base for sound management and policy development.

Consistent quality and timeliness of the science products from cooperative research ensures its place among the core science for meeting management and policy needs. The Federal Information Quality Act (section 515 of the Treasury and General Government Appropriations Act for Fiscal Year 2001, Public Law 106-554) and the Office of Management and Budget's Peer Review Guidelines provide the mandates and guidance on the quality of information used in management. The Information Quality Act issues government-wide guidelines that "provide policy and procedural guidance to Federal agencies for ensuring and maximizing the quality, objectivity, utility, and integrity of information (including statistical information) disseminated by Federal agencies."

The National Oceanic and Atmospheric Administration (NOAA) implements this guidance via the NOAA's Information Quality Guidelines, within which peer review is central to delivery of sound science products and advice. As is true for all science used for management, cooperative research can benefit from peer review at three key levels: (1) programmatic reviews, (2) proposal reviews, and (3) review and integration of final research products.

Programmatic Reviews

An ideal cooperative research program is well planned and comprehensive—one in which funded projects collectively advance the state of knowledge and move us toward a common goal. Conversely, a pitfall to avoid is devolution into a fragmented program—one in which many small and unrelated research projects are funded. While each small project may result in useful findings, the opportunity for synergy is lost in this approach. Programmatic reviews are important tools for setting research priorities—for establishing a balanced research portfolio given present and projected research requirements. Collaborative planning allows for focused research investments that comport with short- and long-term strategic plans and promotes a more comprehensive approach to science in which projects collectively contribute to a growing understanding of a larger problem. These reviews also provide a mechanism for responding to changes in science requirements driven by changes in legislative mandates, management approaches, or policy decisions.

"Consistent quality and timeliness of the science products from cooperative research ensures its place among the core science for meeting management and policy needs."

Another benefit of programmatic reviews is the opportunity they provide to modify science priorities and management decisions based on outcomes from previous research. These feedback loops are crucial to integrating cooperative research and management.

Proposal Reviews

Competition for research funding triggers a peer review process that improves the overall quality of research programs in three ways. First, it creates an opportunity to review the technical merits of the project. A careful review of the study design allows an evaluation of its ability to deliver given the proposed methodology. For example, sample size requirements can be evaluated using power analyses to determine if they are sufficient to meet the desired level of precision. Once a problem is identified, revisions/corrections can be made to flawed methodologies of an otherwise fundable project, allowing that project to go forward.

Second, proposal reviews provide the mechanism for evaluating proposed projects against research plans and established priorities, ensuring that, collectively, selected projects advance us toward larger goals. These evaluations also yield opportunities to leverage projects to improve overall program efficiency.

Cooperative research catalyzes collaborations among sectors with different perspectives and approaches to achieving agreed upon goals. The ideal stage for collaboration is during proposal development, to capitalize on the breadth of experience and knowledge brought to the table by diverse collaborators. The third benefit of strong proposal review processes is the additional opportunity they present for dialog and collaboration among the diverse sectors engaged in or affected by the research. Independently, academic, government, or industry investigators might approach a research question in completely different ways, each with their own strengths. Proper representation of all participating sectors on review panels provides a balanced evaluation and an opportunity to improve projects through feedback coming from diverse expertise.

“Rigorous review of research programs’ products ensures their quality and builds resource managers’ and stakeholders’ confidence in cooperative research programs.”

Review of Science Products

One of the most important steps in finalizing a cooperative research project is carrying out a peer review of the research products. Products from cooperative research range broadly (e.g., raw data, analyses derived from raw data, models and model outputs, new sampling protocols, and gear designs), requiring that selection of the peer review process and reviewers be tailored to the type of product. Living marine resource management is carried out under prescriptive timelines that impact timeliness demands for science inputs. Proper peer review should be built into the project plan and scheduled as part of the delivery of the product to avoid pressure to release or use a product prior to being reviewed.

Cooperative research contributes significantly to our full research capability. Rigorous review of research programs’ products ensures their quality and builds resource managers’ and stakeholders’ confidence in cooperative research programs. In this way, the results of cooperative research can be integrated with other scientific findings into a single stronger scientific basis for decision making.

American Fisheries Society Symposium 52:215–222, 2006

Integrating Cooperative Research and Management

ROBERT S. POMEROY
Department of Agricultural and Resource Economics/Connecticut Sea Grant
University of Connecticut-Avery Point, 380 Marine Science Building
1080 Shennecossett Road, Groton, Connecticut 06340, USA

Comanagement Defined

Cooperative management or comanagement can be defined as a partnership arrangement in which the community of local resource users (fishers), government, other stakeholders (boat owners, fish traders, boat builders, business people, etc.) and external agents (nongovernment organizations [NGOs], academic and research institutions) share the responsibility and authority for the management of the fishery. Through consultations and negotiations, the partners develop a formal agreement on their respective roles, responsibilities, and rights in management, referred to as "negotiated power." Comanagement is also called participatory, joint, stakeholder, multiparty, or collaborative management. Comanagement covers various partnership arrangements and degrees of power sharing and integration of local (informal, traditional, customary) and centralized government management systems (Pomeroy and Rivera-Guieb 2006).

There is no blueprint or model for comanagement, but rather a variety of arrangements from which to choose to suit a specific context. Comanagement should be viewed not as a single strategy to solve all problems of fisheries management, but rather as a process of resource management, maturing, adjusting, and adapting to changing conditions over time. A healthy comanagement process will change over time in response to changes in the level of trust, credibility, legitimacy, and success of the partners and the whole comanagement arrangement. Comanagement involves aspects of democratization, social empowerment, power sharing, and decentralization. Comanagement attempts to overcome the distrust, corruption, fragmentation, and inefficiency of existing fisheries management arrangements through collaboration. Comanagement is adaptive; that is, through a learning process, information is shared among partners, leading to continuous modifications and improvements in management. Through comanagement, the partners actively contribute and work together on fisheries management. They share the costs and benefits and the successes and failures.

Comanagement is not a regulatory technique, although regulations are used in comanagement. It is a participatory management strategy that provides and maintains a forum or structure for action on empowerment, rule making, conflict management, power sharing, social learning, dialog and communication, and development among the partners. Comanagement is a consensus-driven process of recognizing different values, needs, concerns, and interests involved in managing the resource. Partnerships, roles, and responsibilities are pursued, strengthened, and redefined at different times in the comanagement process, depending on the needs and opportunities, the legal environment, the political support, capacities of partners, and trust between partners. The comanagement process may include formal and/or informal organizations of fishers and other stakeholders.

> "Comanagement is adaptive; that is, through a learning process, information is shared among partners, leading to continuous modifications and improvements in management."

The potential advantages of comanagement include the following:

1. A more transparent, accountable, and autonomous management system.
2. A more democratic and participatory system.

3. more economical than centralized management systems, requiring less to be spent on management administration and enforcement, in the long run.
4. Through involvement in management, fishers take responsibility for a number of managerial functions.
5. Makes maximum use of indigenous knowledge and expertise to provide information on the resource base and to complement scientific information for management.
6. Improved stewardship of aquatic and coastal resources and management.
7. Management is accountable to local areas. Fishing communities are able to devise and administer management plans and regulatory measures that are more appropriate to local conditions (localized solutions to local problems).
8. By giving the fishers a sense of ownership over the resource, comanagement provides a powerful incentive for them to view the resource as a long-term asset rather than to discount its future returns.
9. Various interests and stakeholders are brought together to provide a more comprehensive understanding of the resource.
10. Since the community is involved in the formulation and implementation of comanagement measures, a higher degree of acceptability, legitimacy, and compliance to plans and regulations can be expected.
11. Community members can enforce standards of behavior more effectively than bureaucracies can.
12. Increased communication and understanding among all concerned can minimize social conflict and maintain or improve social cohesion in the community.

Despite all these advantages, comanagement has several disadvantages and problems, including the following:

1. It may not be suitable for every fishing community. Many communities may not be willing or able to take on the responsibility of comanagement.
2. Leadership and appropriate local institutions, such as fisher organizations, may not exist within the community to initiate or sustain comanagement efforts.
3. In the short run, there are high initial investments in time, financial resources, and human resources to establish comanagement.
4. For many individuals and communities, the incentive(s)—economic, social, and/or political—to engage in comanagement may not be present.
5. The risks involved in changing fisheries management strategies may be too high for some communities and fishers.
6. The costs for individuals to participate in comanagement strategies (time, money) may outweigh the expected benefits.
7. Sufficient political will may not exist to support comanagement.
8. Unease of political leaders and government officials to share power.
9. The community may not have the capacity to be an effective and equitable governing institution.
10. Actions by user groups outside the immediate community may undermine or destroy the management activities undertaken by the community.
11. Particular local resource characteristics, such as fish migratory patterns, may make it difficult or impossible for the community to manage the resource.
12. The need to develop a consensus from a wide range of interests may lengthen the decision-making process and result in weaker, compromised measures.
13. There may be shifts in power bases (political, economic, social) that are not in the best interests of all partners.
14. There are those who feel that comanagement is too costly and time-consuming and that other alternatives, such as stricter regulations, may be better.

15. There is always a possibility of unbalanced and inequitable sharing of power between government and communities and the use of comanagement by some political leaders solely for their own purposes.

Getting Started: External and Internal Beginnings

Comanagement may have a number of different types of beginnings, that is, the way in which a comanagement program is initiated. The beginnings may be to resolve a resource problem or conflict, to take advantage of a funding opportunity, for political purposes, or to claim resource rights. The beginning of a comanagement program is often highly site and context specific and may involve a variety of different stakeholders. The actual beginning of a program may originate from several different sources and combination of stakeholders and finally evolve into an action. The beginning phase of comanagement may be time-consuming, difficult, and costly. It is not possible to present all the possible scenarios of the beginnings of comanagement program. For the sake of simplicity, two types of beginnings will be discussed. The comanagement program may be (1) initiated external to the community, or (2) initiated internally in the community.

"...it is recommended that community members be consulted and allowed to participate as early in the program life as possible in order to obtain their input, support, and buy-in."

A comanagement program, initiated externally to the community, is one in which the idea for the program originates outside of the community. For example, the program may begin where an external agent (i.e., NGO, academic or research institution) and/or government agency identifies a problem(s) (for example, poverty in fishing communities, overfishing, destructive fishing) plans to address the problem through comanagement, further identifies an area or community in the country to focus the program, and then establishes the program. In another case, the program may be initiated as part of a larger donor-assisted development program in the country in which comanagement is the intervention approach. Or it may be that the government has declared a protected area and wants to develop a comanagement partnership to manage the area.

In the case of a program that begins externally to the community, there may or may not be early consultation and collaboration with the community in designing and preparing for the program. In many externally initiated programs, the details of the program objectives, intervention approach (comanagement), and the specific areas of the program are decided in the design phase away from the community. For example, an NGO might specialize in comanagement and have the resources to undertake a comanagement program in a community. The NGO would then conduct community scoping activities (e.g., a needs assessment) to determine the feasibility of undertaking a comanagement program in the community. Once a program community or area is identified, it is recommended that community members be consulted and allowed to participate as early in the program life as possible in order to obtain their input, support, and buy-in. Although the project idea may not originate with the community, the sooner they are aware that their community has been selected as a program site, the sooner the community members can help shape specific program objectives and strategies. If the program objectives and strategies are kept relatively general at this early stage, community members can be given an opportunity to provide input into further program design and planning and gain a sense of ownership of the project. Program sustainability has been shown to improve when community members are given the opportunity to participate early in the program design and planning stages and have an incentive to want to participate.

Once a community is identified, an externally initiated program may assist community members in problem identification, consensus building, accessing information, and initial action planning. In some cases, an externally initiated program, due to funding or donor demands, may immediately initiate implementation activities such as community organizer integration or education and capacity building.

An internal to the community initiated comanagement program begins from the fishers and other stakeholders in the community. For example, the fishers organize themselves to solve specific conflicts and problems and to claim specific rights. In contrast to the external beginning of a program, this can be considered as a bottom-up beginning for comanagement. This type of beginning may be more promising in terms of sustainability of the program; as the fishers themselves have recognized a problem, they have an incentive to take action and they take the lead to finding a solution. This is not to say that an internal beginning always leads to improved program sustainability as compared to a program with an external beginning. Additionally, to distinguish between an externally initiated and an internally initiated program may be difficult because of highly complex interactions and interlocking relationships and interests between an external agent and the community. For example, a program of the national government may want to expand its operation to a new community that has long recognized their problems but is not able to do something about it. In this case, the idea to initiate some kind of management already exists in the community and this is put into action by an external institution with a similar idea and interest. The important point, whichever type or combination of beginning, is that the fishers and other stakeholders have an incentive (economic, social, political) to want to participate and are given the opportunity to take ownership of the comanagement process early in the project life.

Cooperative Research

There are a number of specific activities and interventions in the implementation phase of comanagement. A major activity in the comanagement process is research.

> The role of research in comanagement is to help establish baselines, inform the management process, and nourish community education and involvement. A common mistake is to focus on research to the exclusion of education and action. By involving community members in these activities, the research process itself becomes one of education and action. In this way, such participatory research lays the foundation of awareness and commitment from which other activities grow. [White et. al 1994]

Information for comanagement will come from scientific studies by experts and from participatory research with resource users and other community members. Scientific information is very useful and important, but the type of information collected by scientists often differs from that obtained from resource users, and the tools and methods of collecting the information are also different (Walters et al. 1998). A significant amount of information can and should come from the community. The indigenous or local knowledge of resource users and other community members (including women and elders) is critical information for planning and management. There are a number of tools and methods available that involve the extensive participation of local community members in gathering and analyzing information and in obtaining indigenous or local knowledge. The combination of scientific and local knowledge can complement each other and greatly enhance the comanagement planning and management program. The collection of information may take several weeks to several months, depending upon the scope and scale of information needs.

“...the fishers and stakeholders have an incentive (economic, social, political) to want to participate and are given the opportunity to take ownership of the comanagement process early in the project life.”

The conventional approach to research has been characterized by control by outside experts, scientists, and development specialists who have set project agendas and carried out research without any or only minor input from local community members (Chambers 1994). Not only have local people not played a part in the planning and implementation of such projects, but their knowledge of local ecology and the structure of their social, economic, and political systems have been ignored as well. The process is relatively static, one in which information is gathered from a community and then processed and analyzed by experts with little or no feedback to the community. Consequently, many projects have failed due to inappropriate project goals, community apathy, and a lack of understanding of local social and ecological systems (Landon and Langill 1998).

In recent years, new approaches to research have been developed that involve community members in gathering information in a participatory manner. Participatory research (PR) represents a family of methodological approaches increasingly accepted and utilized to involve local people in research projects taking place in their own communities. Participatory research is characterized by a cyclical, ongoing process of research, reflection, and action, which seeks to include local people in designing the research, gathering information, analyzing data, and taking action. A key objective of PR is to empower community members by utilizing local knowledge and practices and by giving local people the opportunity to learn skills about and share in the research process. It is also intended to contribute

"...many projects have failed due to inappropriate project goals, community apathy, and a lack of understanding of local social and ecological systems."

Conventional versus participatory research

	Conventional research	Participatory research
Purpose	To collect information for diagnosis, planning, and evaluation	To empower local people to initiate action
Goals of approach	Predetermined, highly specified	Evolving, influx
Approach	Objective, standardize, uniform approach, blueprint to test hypothesis, linear	Flexible, diverse, local adaptation, change encouraged, iterative
Modes of operation	Extractive, distance from subject, focus on information generation	Empower, participatory, focus on human growth
Focus of decision making	External, centralized	Local people, with or without facilitator
Methods/techniques	Highly structured focus, precision of measurement, statistical analysis	Open-ended, visually interactive, sorting, scoring, ranking, drawing
Role of researcher/facilitator	Controller, manipulator, expert, dominant, objective	Catalyst, facilitator, visible initially, later invisible
Role of local people	Sample, targets, respondent passive, reactive	Generators of knowledge, participants active, creative
Ownership of results	Results owned and controlled by outsiders, who may limit access by others	Results owned by local people, new knowledge resides in people
Output	Reports, publications, possible policy change	Enhanced local action and capacity, local learning, cumulative effect on policy change, results may not be recorded

Source: Narayan 1996

directly to positive changes in the specific circumstances of the participants, as well as increase the chances that the comanagement program will succeed through local involvement (Landon and Langill 1998).

It is important to note that the use of participatory research does not mean that conventional scientific research conducted by outside experts should not be conducted. There are a number of circumstances where this will be the most appropriate method of gathering information. There is overlap and complementaries between the two approaches to research. For example, participatory research may identify an issue, which is then studied in more depth through conventional research.

Conventional research is most appropriate

1. when data needed are mostly quantitative;
2. when follow up action, in terms of program and project training, is uncertain;
3. when issues addressed are not sensitive;
4. when the purpose of the study does not include setting the stage for staff or community involvement in a program; and
5. when time and resources are not serious considerations.

Participatory research is most appropriate

1. to establish rapport and a commitment to use study results;
2. when staff or community interest and involvement is central to achieving program goals;
3. when information is sensitive;
4. when major issues are unknown or relatively undefined; and
5. when supporting local capacity is important (Narayan 1996).

Among the many approaches that make up participatory research, rapid rural appraisal (RRA) and participatory rural appraisal (PRA) offer the best alternatives to the conventional approach to research. Although there are a large number of descriptions, the most useful are

Rapid rural appraisal emphasizes the importance of learning rapidly and directly from people. Rapid rural appraisal involves tapping local knowledge and gaining information and insight from local people using a range of interactive tools and methods (Jackson and Ingles 1995). Rapid rural appraisal is often used to inform those from outside the community.

Participatory rural appraisal involves field workers learning with local people with the aim of facilitating local capacity to analyze, plan, resolve conflicts, take action, and monitor and evaluate according to a local agenda (Jackson and Ingles 1995).

Rapid rural appraisal is regarded as a set of guidelines and tools that can be used in many different ways and many different circumstances without necessarily attempting to change political and social structures. Participatory rural appraisal is used to specifically refer to a use of RRA approaches and tools to encourage participation in decision making and planning by people who are usually excluded.

Both approaches are carried out by multidisciplinary teams and differ from conventional information-gathering approaches in that field workers work with and learn directly from local people. The methods involve a minimum of outsider interference or involvement.

While the speed with which data can be collected using these methods is an advantage, it also introduces the risk of accepting superficial descriptions of situations and adopting superficial solutions to issues. Rapid rural appraisal and PRA are often seen as a replacement for other forms of investigation and study even in situations where more formal or analytical research is called for. These methods can be biased towards local people who have time and motivation to talk to field workers and towards people who appear to have knowledge. When facilitated in highly formalized fashion or applied too rigidly and repetitiously, their potential effectiveness is reduced. Good practitioners will use a variety of methods for cross-checking information and the applicability of proposed solutions. The depth of understanding of issues should be undertaken over time and with other methods.

It should be noted that most community members have limited experience with participation and participatory research methods and time will need to be allocated to empower community members to actively participate in PRA activities. In addition, PRA should not be thought of as merely about tools that would lead to some initial form of community analysis about their problems and issues. Participatory rural appraisal, more than anything else, is about the shared experience of the researchers and communities in a cycle of reflection and action. It is not just doing research with local people; it is part of a much broader goal on people empowerment (Brzeski et al. 2001).

> "...most community members have limited experience with participation and participatory research methods and time will need to be allocated to empower community members to actively participate in PRA activities."

Good RRA and PRA are characterized by behavior and attitudes that build rapport with local people, including avoiding putting people in uncomfortable situations, learning from people not lecturing to them, creativity, checking and rechecking the validity of the information obtained, listening and probing, being patient, engaging in conversations that have two-way exchange of information, being trustworthy, and being open and friendly (Jackson and Ingles 1995).

Research Opportunities

A number of hypotheses exist concerning the advantages of comanagement, especially as comanagement compares to centralized management. These hypotheses include

- legitimacy: increased authority of the organization, regulations, management system;
- transaction costs: overall reduction;
- incentives: behavior modification;
- enforcement and compliance: improvement;
- institutional resiliency: flexibility and adaption of institutional and organizational arrangements;
- conflict management: improvement;
- ecosystem health: biological improvements;
- rationality: increased knowledge, data and information sharing;
- democracy (empowerment, representation, choice): reduced political and equity problems; and
- stewardship: improvement.

There are a number of research questions related to the process of comanagement and comanagement systems:

- Organizational form: most appropriate and effective
- Scale: of institutional and organizational arrangements, ecosystem, user
- Process: institutional and organizational arrangements evolution over time
- Structure and content of comanagement agreements
- Impact: measurement of short- and long-term impacts

References

Brzeski, V., J. Graham, and G. Newkirk. 2001. Participatory research and CBCRM: in context. Coastal Resources Research Network, Dalhousie University, Halifax, Nova Scotia and International Development Research Centre, Ottawa.

Chambers, R. 1994. Participatory rural appraisal (PRA): analysis of experience. World Development 22(9):1253–1268, Montreal.

Jackson, W. J., and A. W. Ingles. 1995. Participatory techniques for community forestry: a field manual. ANUTECH Pty. Ltd., Nepal-Australia Community Forestry Project, Technical Note 5/95, Canberra, Australia.

Landon, S., and S. Langill. 1998. Participatory research: readings and resources for community-based natural resource management researchers. Volume 3. International Development Research Centre, Ottawa.

Narayan, D. 1996. What is participatory research? Pages 17–30 *in* Toward participatory research. World Bank, Washington, D.C.

Pomeroy, R. S., and R. Rivera-Guieb. 2006. Fishery co-management: a practical handbook. CABI Publishing, Cambridge, Massachusetts and International Development Research Centre, Ottawa.

Walters, J. S., J. Maragos, S. Siar, and A. White. 1998. Participatory coastal resource assessment: a handbook for community workers and coastal resource managers. Coastal Resource Management Project and Silliman University, Cebu City, Philippines.

White. A., L. Z. Hale, Y. Renard, and L. Cortesi. 1994. Collaborative and community-based management of coral reefs. Kumarian Press, West Hartford, Connecticut.

American Fisheries Society Symposium 52:223–225, 2006

Integrating Cooperative Research and Management: Perspectives from a Recreational Fishing Organization

Jason Schratwieser
International Game Fish Association
300 Gulf Stream Way, Dania Beach, Florida 33004, USA

Introduction

The International Game Fish Association (IGFA) was founded in 1939 and is a not-for-profit organization committed to the conservation of game fish and the promotion of responsible, ethical angling practices through science, education, rule making, and record keeping. Originally housed in the American Museum of Natural History in New York City, IGFA has always had strong ties with fisheries research. Indeed, the appointment of William King Gregory, Ph.D., who at the time was chairman for both the ichthyology and comparative anatomy departments, as IGFA's first president solidified the organization's dedication to conservation and increasing the world's knowledge of fishes.

In the subsequent 66 years, IGFA has endeavored in its goal of promoting the sport of angling not only as recreation, but also as a source of scientific data. In addition to compiling decades of catch information from around the world, IGFA staff, trustees, and international representatives have participated in international cooperative research and management efforts. Presently, IGFA represents its membership and recreational anglers, in general, on numerous regional, national, and international fisheries management panels and also funds and participates in research relating to game fish and their habitats.

Whether participating in cooperative research or management, IGFA's core purpose is to provide a link and facilitate interaction and information exchange between recreational anglers and fisheries scientists and managers. Thus, the following are perspectives on the concept of integrating cooperative research and management from the position of an organization acting in the liaison capacity.

How Do Cooperative Research and Management Guide Each Other?

One of the first things to consider when approaching this subject is that cooperative research and cooperative management are distinct components of a single, bigger process. One of the inherent challenges to cooperative management is that the people involved often have diverse backgrounds not associated with science or research, which often leads to a fundamental distrust or misunderstanding of the science that is driving management. This produces negative results, including unproductive discussions between stakeholders and managers, diminished compliance with new management decisions, and reduced likelihood for future participation in cooperative management.

> "...to provide a link and facilitate interaction and information exchange between recreational anglers and fisheries scientists and managers."

One way to combat the issue is to involve stakeholder groups earlier in the process by getting them to participate in cooperative research, when and where appropriate. Properly implemented cooperative research can generate stakeholder buy-in, improve compliance with management decisions, and serve to perpetuate a working relationship between stakeholders and management.

The concept of involving stakeholders in cooperative research may appear as somewhat of an oversimplification. However, the difficultly lies in the details of how and when to involve stakeholders. Caution must be exercised when doing so as there are several pitfalls that can jeopardize the entire process. First and foremost, a cooperative research effort involving stakeholders should only be initiated on the premise that there is a clear objective to be met and that the results can potentially affect management or benefit the resource. That is, cooperative research should not be done solely as a gesture of demonstrating cooperation. Furthermore, the data collected from these studies should not be viewed as second-class science.

Another important aspect to consider is that specific stakeholders should be involved because they provide assets or skill sets that are complementary to governmental or academic scientists that will improve the quality of research. It is unrealistic, for example, that recreational anglers will be able perform tasks such as stock assessments or spatial analyses. However, they can be beneficial in locating and catching fish for fisheries dependent data, tagging studies, or experimental gear research.

Example: Circle Hook Research

One of the biggest benefits of involving recreational anglers in cooperative research is that they can help supply sample sizes that otherwise might be unobtainable or take an unfeasible amount of time. Recently, the IGFA and the National Marine Fisheries Service conducted a study comparing several types of hooks on the recreational sailfish *Istiophorus platypterus* fishery in south Florida. The objective of the study was to determine which type of hook provided the best chance of postrelease survival and to see how this varied by different drop back times or the time given for the fish to eat the bait before engaging the hook.

Statistically evaluating three different hook types over four drop back categories required that data be collected from a considerable number of sailfish. Sailfish also happen to be a fish that requires specialized equipment and techniques to catch in appreciable numbers. Enlisting the help of only nine captains over two sailfish seasons provided data on more than 2,000 sailfish. This sample size was sufficient to make statistical inference on the different hooks' performance over the four different drop back categories.

It is worth mentioning that the study found that using a particular model of circle hook greatly reduced the occurrence of gut hooking, while providing identical catch rates to other hooks tested in the study. Many recreational anglers have been reticent to use circle hooks because, even though they know they benefit fish, there continues to be a perception that they are not as effective at hooking and catching fish. Thus, the benefit of incorporating anglers in this study is actually twofold. First, they supplied a sample size large enough to statistically demonstrate that circle hooks are beneficial to fish without compromising catch rates. Second, because we used the best sailfish captains and anglers in south Florida to collect the data, anglers in general will be more likely to accept the results of the study as credible and adopt the use of these hooks because the data were collected by their peers.

> "...a cooperative research effort involving stakeholders should only be initiated on the premise that there is a clear objective to be met and that the results can potentially affect management or benefit the resource."

Initiating Cooperative Research

As previously discussed, cooperative research involving anglers or other stakeholders should be initiated on the basis of completing clear objectives. Thus, the identification of a study question and objectives should be a first step, after which appropriate stakeholders are solicited for participation based on skills or assets that they can provide.

Soliciting the help of stakeholders presents its own set of challenges. For example, recreational anglers are often apprehensive about participating in projects that involve state or federal fisheries management. The reasons for this vary but generally stem from a history of poor communication and distrust. In addition, fisheries management may find the process of identifying and contacting individual stakeholders difficult and inefficient.

In scenarios such as this, a better solution is for management to identify and contact entities that represent specific stakeholder groups. Nongovernmental organizations (NGOs) and advocacy groups can help facilitate cooperative research in several ways. First, they provide management with a single contact and source of communication to stakeholders. Organizations like IGFA that have staff biologists with experience in fisheries management also make communication more efficient between management and stakeholders by disseminating technical information in a more comprehensible fashion. Nongovernmental organizations and advocacy groups are better suited to identifying stakeholders that would best aid the research in question. Finally, they can also aid in publicizing results of cooperative research, which can further cultivate relations between stakeholders and management for future cooperative efforts.

“Nongovernmental organizations and advocacy groups are better suited to identifying stakeholders that would best aid the research in question. Finally, they can also aid in publicizing results of cooperative research, which can further cultivate relations between stakeholders and management for future cooperative efforts.”

Conclusions

Integrating cooperative research and management means understanding that research and management are distinct entities of a bigger process. Involving stakeholders in cooperative research can lead to more productive management because there is better understanding of the science driving management policy. However, cooperative research involving the public is not a magic bullet to improve relations with management. Clear objectives that can potentially influence management policy or directly benefit the resource are necessary for stakeholder involvement. Stakeholders should be solicited for participation because their assistance will improve the quality of research. Finally, NGOs and advocacy groups can facilitate much more efficient cooperation by acting as liaisons between management and stakeholders.

American Fisheries Society Symposium 52:227–231, 2006

Developing a Regional Research Partnership in the Columbia River Basin

STEVE M. WASTE
Northwest Power and Conservation Council
851 S.W. Sixth Avenue, Suite 1100, Portland, Oregon 97204, USA

The Northwest Power and Conservation Council (Council) Columbia River Basin Fish and Wildlife Program is one of the largest regional efforts in the nation to recover, rebuild, and mitigate impacts of hydropower dams on fish and wildlife. As a planning, policy-making, and reviewing body, the Council develops and then monitors implementation of the program, which is funded by the Bonneville Power Administration and implemented by tribal, state, and federal fish and wildlife managers and others.

For over 20 years the Council has supported a diverse range of research efforts. Hundreds of excellent projects, including dedicated research projects and restoration projects with research elements, have been completed since the inception of the program in 1982. Projects implemented under the Council's program and others in the Columbia River basin have substantially advanced the state of scientific understanding of fish and wildlife restoration. Yet the continuing absence of a plan to coordinate research has contributed to a lack of focus on key research needs. To complement its traditionally strong support for research, the Council has approved a Columbia River basin research plan to guide the development of a research under its program and to coordinate with the research programs of other entities within the region.

"The research plan will reduce management uncertainty by increasing scientifically based knowledge."

The Columbia River basin research plan recognizes other research plans as important components of a potentially integrated regional research program and provides a framework for establishing linkages between existing research programs and initiatives. This plan recommends research to be funded through the program, as well as recommendations for research that will require collaborative, multiparty funding commitments by the Council and other entities with similar research mandates.

How Do Cooperative Research and Management Inform and Guide each Other? What Opportunities Exist for the Future?

Research reduces management uncertainty

The research plan will reduce management uncertainty by increasing scientifically based knowledge. Research is necessary to provide scientifically credible answers to management questions arising from key uncertainties. The research plan profiles critical uncertainties and research recommendations identified by the Council's scientific review groups, fish and wildlife managers, and other agencies and entities within the Columbia River basin. Several of the research questions in the region have persisted over many years because of the inability of resource management agencies to secure or collaborate on the funding necessary to mount large-scale field experiments to address them. By reducing uncertainty, course corrections can be achieved that will increase the efficiency and the effectiveness of the program. The research plan splits complex issues into treatable questions, with the expectation that answers to one question may shed light and help resolve related questions. Plan objectives include

1. identification of key uncertainties and research recommendations;
2. prioritization of major research topics;

3. accountability for the annual expenditures of research funds;
4. input from independent scientific review, fish and wildlife agencies and tribes, independent scientists, and other interested parties in the region;
5. monitoring, evaluation, and the application of results;
6. coordination with the research elements of the main-stem plan;
7. coordination with the research elements of the subbasin plans; and
8. making information from the Fish and Wildlife Program readily available.

The plan is an important tool for managing the program because it is designed to inform Council decision making, facilitate scientific review, focus project selection, and provide a basis for redirecting research in the future. The plan also explains the Council's basic principles for the selection, implementation, completion, and processing of research conducted under its program.

Cooperation: a catalytic ingredient

In the 2000 Fish and Wildlife Program, the Council called for a reevaluation of our current approach to research and a plan to guide future research. Many other resource management entities share responsibility with the Council for conducting within the Columbia River basin, and the Council recognizes that the status quo is composed of several separate research plans. These plans make general reference to the "need to coordinate" with other similar efforts, but rarely set forth any explicit steps to implement such coordination. Consequently, staff developed a plan that does more than just meeting the requirements of the program; it also facilitates integration between the disparate research efforts within the region.

The previous lack of guidance for basin-wide research has hampered our ability to focus on key research needs. Looking back, it is clear that difficulty agreeing on problem definition coupled with shared funding responsibilities under overlapping mandates have resulted in a fragmentation of effort that explains why key uncertainties have persisted. Clearly, cooperation is the catalytic ingredient required to develop a cohesive approach to research within the Columbia River basin.

What Are Future Research Opportunities, Needs, and Issues Ecosystem Approaches, Better Data, Habitat Data, Geographical Data, and Protocol Issues (e.g., Problems Involved in Collecting the Data, New Methods to Spatially Analyze Fisheries Data)?

Identifying priorities for future research: developing a regional research agenda

The critical uncertainties identified in the research plan were generated from evaluation of independent science group reports, recommendations from national science groups, the Fish and Wildlife Program, the biological opinions, and other regional research plans. These were organized into 13 research topics that varied in terms of their magnitude. Some topics were dominant in terms of current activity, such as hydrosystem and hatchery effectiveness. Research activity in other areas was modest, or nascent in the case of emerging issues such as toxics or invasive species. Many parties contributed to the development of the plan through a public review process. The plan was developed with the input of tribes, states, and federal agencies. Key partners in this endeavor include the Columbia Basin Fish and Wildlife Authority (CBFWA) and the Columbia River Intertribal Fish Commission, who are also presenting in this symposium.

"The previous lack of guidance for basin-wide research has hampered our ability to focus on key research needs."

An approach to implementation: inaugurating a regional research partnership

To achieve cooperative management in support of research within the Columbia River basin, the plan recommends the convocation of a forum for research program leads (i.e., a research partnership). The region currently lacks a forum where researchers can cross disciplinary and institutional boundaries and find peer support for potentially controversial recommendations. Prior attempts have been made to convene executive level multi-agency groups and fora for the purpose of coordinating resource management decision making across the Columbia River basin.

A key challenge for the research partnership is to move beyond the piecemeal solutions that have undercut the overall success of prior restoration efforts and design a comprehensive effort to reduce sources of mortality across the life cycle of the salmon. The plan recommends that policy makers such as the Council members and other regional executives foster integration of the currently compartmentalized research agendas and budgets of entities that share common objectives by identifying shared research priorities and developing collaborative implementation strategies.

Several meetings were held with potential partners to discuss these points, with CBFWA in Portland and with the National Oceanic and Atmospheric Administration, the U.S. Geological Survey, and the U.S. Environmental Protection Agency in Seattle. The comments indicated strong support for the Council's research plan as a document that facilitates a regional approach to research and also inaugurates a process for coordinating existing research initiatives.

Future opportunities

In the past, attempts have been made to convene executive level multi-agency groups and fora for the purpose of coordinating resource management decision making across the Columbia River basin. These unsuccessful efforts indicate that it may be not be possible to convene a single "super-group" that can address management decisions across all subject matter areas of resource management in the Columbia River basin. This plan proposes an alternative approach that would provide bookends to this spectrum of resource management issues. This would be achieved through the convocation of a consortium of three separate, collaborative partnerships. One would be a research partnership able to transcend the institutional impediments described in the preceding section. Many of the resource management entities contacted during the development of this research plan expressed support for this concept. The second would be a data management partnership, a concept for which Council-sponsored projects and support already provide significant substance. The third would be a partnership for monitoring, which already exists in the form of the Pacific Northwest Aquatic Monitoring Partnership. This group would provide the feedback mechanism that is missing at the programmatic scale.

“A key challenge for the research partnership is to move beyond the piecemeal solutions that have undercut the overall success of prior restoration efforts and design a comprehensive effort.”

If such a configuration of partnerships were to become viable, the research partnership would increase the ability of the region to reduce scientific uncertainty, the data partnership would provide a repository for analytical manipulation, and the monitoring partnership would support the evaluation that has long been missing from the Columbia River basin. Even if the three partnerships were only semiformal in an administrative sense, and only loosely coupled in a decision-making sense, the synergy that would result could significantly increase the ability of the region to redirect its efforts based on ongoing experience.

Cooperatively managing monitoring and evaluation at a regional scale

A key challenge is how to evaluate research and restoration efforts at different scales in a way that is scientifically defensible and ecologically meaningful (e.g., how to link monitoring efforts at the watershed or subbasin scale with efforts at the larger scale of evolutionarily significant units). To address these monitoring needs in the Columbia River basin in any practical sense requires a shift from work at the project scale to a programmatic approach. Individual subbasins cannot be expected to have or hire the specialized expertise, secure long-range funding commitments, or command the ability to mount and sustain long-term monitoring and evaluation efforts. Yet, because the issues of interest to subbasin planners in monitoring and evaluation are also of interest to entities with responsibilities over a broader geographic scale, it is possible for efforts at the smaller scale to benefit from the overlay of efforts at a broader scale. Therefore, rather than try to design a complete and comprehensive monitoring program within each subbasin, which it probably cannot afford, the region should identify and develop consensus about how much and what type of monitoring is needed and can be afforded for managing an effective program.

“Sound science, thoughtful planning, and hard work on the ground are all important ingredients to the success of any restoration program.”

For these reasons, the Council supports the efforts of the Pacific Northwest Aquatic monitoring program (PNAMP) to develop a regional approach to monitoring. The PNAMP is a voluntary, non-directive, self-organizing forum committed to developing a unified approach to monitoring in the Pacific Northwest. The PNAMP provides a forum for collaboration of the members as they work towards developing a coordinated approach to monitoring by providing a durable structure for facilitating the development of cross-party linkages. Professional courtesy and respect for overlapping but different mandates is an essential ingredient for the successful negotiation of a regional approach to monitoring in the Pacific Northwest.

How to Get Cooperative Research or Cooperative Management Started in a New Place? Should It Be Implemented in Other Fisheries Contexts?

In the Columbia River basin, three large-scale planning initiatives are currently underway: planning for implementation of the Federal Columbia River Power System Biological Opinion, recovery planning, and subbasin planning. The NPCC's subbasin planning initiative is an exercise in cooperative planning that addresses management and research issues with many entities such as the tribes, fishing industry, and environmental groups.

Subbasin plans: a new start at a different scale

Sound science, thoughtful planning, and hard work on the ground are all important ingredients to the success of any restoration program. Yet, without strong local support for restoration activities, the future of many Columbia River basin species will remain in question. Therefore, in 2000, the Council initiated subbasin planning to help local entities develop their own restoration plans. Subbasin planning has helped people define the future they seek for natural resource values in their subbasins and thereby define their legacy to future generations.

Subbasin plans have identified coordination needs and opportunities for fish and wildlife restoration by integrating strategies in the program with other federal, state, tribal, Canadian, and volunteer fish and wildlife restoration programs. The cooperative and inclusive participation of diverse stakeholders has enabled subbasin plans to contain a collective expression of the critical uncertainties and research priorities. Many subbasin plans have identified research needs that are either site specific or a prevalent need within the subbasin or province. The research plan supports research recommendations that have broad application to other provinces or to the entire Columbia basin.

Recovery planning: balancing curative with preventative approaches

Today, the planning arena within which the fish and wildlife restoration is being implemented is in transition, as the region is reconfiguring the program to address new responsibilities under the Endangered Species Act (ESA). The region must emphasize a balanced approach to implementing the Fish and Wildlife Program, despite strong external pressures to shift the entire program into supporting implementation of ESA. Shifting program emphasis too far in the direction of the ESA could become self-defeating, as the curative approach embodied in the ESA is expensive and the outcomes are uncertain. In contrast, the program embodies the preventative approach of protecting the viability of all affected species to preclude additional listings under the ESA. The preventative approach is less expensive and more likely to protect existing fish and wildlife.

How to get started: enfranchise future partners

A formal public comment period on the draft Columbia River basin research plan was held from October 1 to October 30, 2004. A total of 28 comments were received from the tribes (three), state agencies (eight), federal agencies (eight), local governments (one), academic institutions (two), consulting firms (four), and private individuals (two). The types of comments received ranged from very general points affecting the organization of the document to very specific comments on a particular research topic. After revisions in response to public comments, the plan was submitted it to the Council's Independent Scientific Advisory Board and Independent Scientific Review Panel for review.

> "The preventative approach is less expensive and more likely to protect existing fish and wildlife."

Collaboration: the key to success

The plan recommends that a research partnership be convened to provide a forum for the identification of shared research priorities and development of collaborative implementation strategies. Widespread support for this concept was expressed through the public comments, which recommended that such a partnership collaboratively identify regional research priorities, facilitate implementation of the plan through cost-sharing and other means, and, develop means for the dissemination of research results. The CBFWA offered to continue working with Council staff through the revision process and to help sponsor the regional research partnership.

Part III: Symposium Poster Abstracts

Editors' Note:

The Symposium Organizing Committee solicited posters to be presented as part of the program. The addition of these posters added a great deal of diversity and richness to the overall symposium content, since they detailed specific cases of cooperative fisheries research or management. During the panel presentations and discussions, we discussed broader impacts, lessons learned, and opportunities for the future in cooperative research and management, rather than any particular example. Given this program design, there was not an opportunity to present empirical research and detailed case history overviews, but showcasing individual examples of cooperative research and management programs would promote greater networking among participants and extend the discussion of the themes from sessions. The posters highlighted some of these important and innovative cooperative research and management programs. There was an accompanying reception to facilitate networking, dialog, and learning.

American Fisheries Society Symposium 52:235–247, 2006

Symposium Poster Abstracts

Abstract Title: Sea Grant Fisheries Extension Activities in North Carolina, South Carolina, Georgia, and Florida
Authors: Baker, Jr., M.; Adams, Charles; Vendetti, Jr., Richard; Von Harten, Amber

M. Scott Baker, Jr.
North Carolina Sea Grant Extension Program
UNC-W Center for Marine Science
5600 Marvin Moss Lane
Wilmington, North Carolina 28409, USA

Charles M. Adams
University of Florida
Florida Sea Grant Extension Program
Post Office Box 110240
Gainesville, Florida 32611, USA

Richard A. Vendetti, Jr.
University of Georgia Marine Extension Service
715 Bay Street
Brunswick, Georgia 31520, USA

Amber Von Harten
South Carolina Sea Grant Consortium
Post Office Box 189
Beaufort, South Carolina 29901, USA

Sea Grant programs in the southeast Atlantic region, from North Carolina through the east coast of Florida, are currently engaged in fisheries extension activities that address a host of topics concerning a wide variety of user groups that include commercial fishermen, recreational anglers, and the interested public. More often than not, individual Sea Grant programs develop their own localized extension programming. There may be instances where regional collaboration in extension programming is warranted. To that end, a collaborative effort has been made to inventory current fisheries extension activities by state, identify areas of commonality in the southeast region, and improve communication between extension programs. Issues of mutual concern within the region for recreational anglers include ethical angling (i.e. use of release methods to reduce release mortality), stock enhancement, and general fisheries management education. Issues of mutual concern for commercial fishermen include providing assistance to the commercial shrimp industry (i.e., bycatch reduction via turtle excluder devices and bycatch reduction devices, Trade Adjustment Assistance Program education, vessel-level seafood quality training, and general education/outreach concerning changing fisheries management policy and regulations). In contrast, there are some educational programs and outreach activities that are specific to each state and do not lend themselves to regional efforts.

Abstract Title: The Pacific Northwest Aquatic Monitoring Partnership: A Forum for Regional Coordination
Author: Bayer, Jennifer

Jennifer M. Bayer
U.S. Geological Survey-Columbia River Research Library
5501A Cook-Underwood Road
Cook, Washington 98605, USA

The purpose of the Pacific Northwest Aquatic Monitoring Partnership (PNAMP) is to provide a forum for coordinating state, federal, and tribal aquatic habitat and salmonid monitoring programs. Improved communication, shared resources and data, and compatible monitoring efforts provide increased scientific credibility, cost-effective use of limited funds, and greater accountability to stakeholders. The PNAMP provides leadership through the development and the advancement of recommendations and agency level agreements that are considered for adoption by the participating agencies. PNAMP has adopted the following goals: (1) improve communication between monitoring programs across state, tribal, and federal organizations; (2) improve scientific information needed to inform resource policy and management questions and decisions; (3) seek efficiencies and cost-effectiveness across monitoring programs through compatible and cooperative monitoring efforts; (4) promote science-based credibility of monitoring and assessment efforts; and (5) share resources and information between monitoring programs across state, tribal, and federal organizations. The PNAMP receives significant policy support and direction by member organizations, commitments of technical resources and staff time, and funding for the coordination itself. As part of a monitoring coordination structure, the PNAMP has identified and developed technical working groups for five key elements of monitoring: watershed condition monitoring, effectiveness monitoring, fish population monitoring, estuary monitoring, and data management.

Kenneth Beal
National Marine Fisheries Service
549 Washington Street
Gloucester, Massachusetts 01930, USA

Jennifer Stritzel-Thomson
Massachusetts Division of Marine Fisheries
30 Emerson Avenue
Gloucester, Massachusetts 01930, USA

Abstract Title: Cooperative Research in the Northeastern USA under the Saltonstall-Kennedy Program
Authors: Beal, Kenneth; Stritzel-Thomson, Jennifer

The Saltonstall-Kennedy Act was passed in 1954, although the competitive grant program, which focused federal funding on regional fisheries issues, has flourished only during the last 25 years. The Northeast Region of the National Marine Fisheries Service has funded more than 250 projects worth about $30 million during this 25-year period. The majority of these projects have been cooperative endeavors with fishermen and fisheries organizations working with biologists from state fishery agencies or universities. The authors will present case studies of several successful cooperative research partnerships focused on innovative gear, new fisheries, new products, higher quality seafood, and better data for fisheries management.

Jennifer A. Bloeser
Pacific Marine Conservation Council
Post Office Box 327
Arcata, California 95521, USA

Caroline F. Gibson
Pacific Marine Conservation Council
Post Office Box 794
Port Townsend, Washington 98368, USA

Astrid J. Scholz
Sarah C. Klain
Ecotrust
721 NW Ninth Avenue, Suite 200
Portland, Oregon 97209, USA

Abstract Title: Spatial COmmunity Outreach Project (SCOOP)
Authors: Bloeser, Jennifer; Gibson, Caroline; Scholz, Astrid; Klain, Sarah

The Spatial COmmunity Outreach Project (SCOOP) is a cooperative effort between the Pacific Marine Conservation Council (PMCC), Ecotrust, and members of the West Coast fishing industry, in consultation with the Port Liaison Project. The Port Liaison Project provided financial compensation to the fishermen participating in SCOOP. The primary goals of this project are (1) to gather information from fishing communities on the impacts of the use of spatial management in the West Coast groundfish fishery, collect community-based recommendations on the application of this management tool, and (2) collect information for improving a coast-wide area model of the groundfish fishery. The principal investigators of this project from PMCC and Ecotrust gathered written and digital mapping information in late summer and early fall of 2004 on the effects of the upper continental shelf closures from 2002 to 2004. Information was collected during 28 interviews of members of the fishing industry in 11 coastal communities, ranging geographically from Eureka, California north to Port Townsend, Washington. The information gathered is now being processed and analyzed and will be presented to the Pacific Fishery Management Council (PFMC) by PMCC and Ecotrust in Spring 2005.

Jennifer A. Bloeser
Pacific Marine Conservation Council
Post Office Box 327
Arcata, California 95521, USA

Flaxen D. L. Conway
Oregon Sea Grant
Oregon State University
307 Fairbanks Hall
Corvallis, Oregon 97331, USA

M. Elizabeth Clarke
NOAA's National Marine Fisheries Service
Northwest Fisheries Science Center
2725 Montlake Boulevard East
Seattle, Washington 98112, USA

Abstract Title: Using the World Wide Web to Inform, Connect, and Support Cooperative Research Efforts on the West Coast
Authors: Bloeser, Jennifer; Conway, Flaxen; Clarke, Elizabeth

Cooperation requires connection. Scientists and fishermen interested in supporting cooperative research efforts on the West Coast have turned to the Internet for some tools to connect people spanning the states of Washington, Oregon, and California. Through the development of cooperative research Web sites, the Internet is being used to facilitate much needed connections between the fishing and science communities for the purpose of increasing and improving cooperative research opportunities. FishresearchWest.org is a neutral, online clearinghouse for information on cooperative fisheries research and provides a platform for formally merging the skills and knowledge of scientists and fishermen along the West Coast to improve fisheries science. The site was developed by the Pacific Marine Conservation Council through funding from the NOAA Fisheries Northwest Fisheries Science Center (NWFSC). Heads Up! (www.heads-up.net) is a collaborative site devoted to connecting all interested in West Coast commercial fishing—industry groups, fishermen/families, community organizations, agencies, and others. This electronic bulletin board delivers concise, timely information helpful in conducting business, making good decisions, and participating in management or other industry issues. Visitors view and share timely news, events, publications, crew needs, and so forth. The site was developed by Oregon Sea Grant with funding from industry and agency partners, including NWFSC.

Abstract Title: Surfclam Management Advice Generated through Partnering of Academia, Government, and Industry
Authors: Bochenek, Eleanor; Wallace, David; Powell, Eric; Weinberg, James

Eleanor A. Bochenek
Eric N. Powell
Haskin Shellfish Research Laboratory
Rutgers University
1636 Delaware Avenue
Cape May, New Jersey 08204, USA

David H. Wallace
North Atlantic Clam Association
Wallace & Associates
1142 Hudson Road
Cambridge, Maryland 21613, USA

James R. Weinberg
National Marine Fisheries Service
Northeast Fisheries Science Center
166 Water Street
Woods Hole, Massachusetts 02543, USA

The surfclam fishery, managed by individual transferable quotas, is a valuable and well-managed exclusive economic zone (EEZ) fishery on the northeast U.S. coast. Through 1999, surfclam stock biomass was fairly stable throughout the mid-Atlantic region. The 2002 National Marine Fisheries Service (NMFS) surfclam survey revealed a reduction in surfclam abundance in the Delmarva region, near the southern limit of the species range. The Delmarva population had not been overfished, and commercial landings were not high enough there to account for the decline. In 2004, the surfclam industry funded a surfclam survey in partnership with academia and government. NMFS assisted with design of the survey, fieldwork, data management, and final analysis. Academia assisted with survey design, fieldwork, and data analysis. Funding for the survey, fishing vessel, and crew was provided by the surfclam industry. Study objectives were to estimate temporal changes in mid-Atlantic surfclam biomass, to evaluate recruitment into the fishery, and to assess changes in the bathymetric distribution of surfclams. Survey results indicate that the mortality event (1999–2002 period) seems to have run its course. Data collected from this cooperative survey will be used by the Mid-Atlantic Fishery Management Council and NMFS to help determine future surfclam quotas for the EEZ and in planning future NMFS surfclam surveys.

Abstract Title: Cooperation: A Critical Component to Improved Science for Fisheries Management
Authors: Clarke, M. Elizabeth; Harms, John; Conway, Flaxen; Bloeser, Jennifer

M. Elizabeth Clarke
John Harms
NOAA Fisheries
Northwest Fisheries Science Center
2725 Montlake Boulevard East
Seattle, Washington 98112, USA

Flaxen D. L. Conway
Oregon Sea Grant
Oregon State University
307 Fairbanks Hall
Corvallis, Oregon 97331, USA

Jennifer A. Bloeser
Pacific Marine Conservation Council
Post Office Box 327
Arcata, California 95521, USA

Reflecting NOAA Fisheries' increased focus on cooperative research, the Northwest Fisheries Science Center (NWFSC) and the West Coast fishing community are cooperating in innovative ways to improve the science used for managing groundfish. NWFSC's cooperative research program is a mosaic of partnerships and opportunities for industry to participate in the research process. A port liaison project was developed to involve individual fishers in research with academia as well as government. This project works through industry leaders in several west coast communities to facilitate the matching of fishers with both agency and university researchers. Since 1998, the center has conducted multiple surveys using chartered fishing vessels to monitor ecologically and economically important groundfish populations. Due in part to the successful cooperation achieved during these surveys, the industry now collaborates with the center in the design of new surveys. The center, in partnership with Pacific States Marine Fisheries Commission, created a program to fund industry initiated research proposals. This program emphasizes industry expertise in all phases of research—from design to completion. These three components are linked by outreach efforts featuring a Web site that is a clearinghouse for West Coast cooperative research. This mosaic leverages resources and fosters collaborations that have improved research, strengthened relationships, and promoted the concept of cooperative fisheries research in general.

Abstract Title: Taking Local Initiative and Building Capacity through Cooperative Fisheries Research in an Historic Oregon Fishing Community
Authors: Cobb, Leesa; Gibson, Caroline

Leesa D. Cobb
Port Orford Ocean Resource Team
Post Office Box 679
Port Orford, Oregon 97465, USA

Caroline Gibson
Pacific Marine Conservation Council
Post Office Box 794
Port Townsend, Washington 98368, USA

Working with several partners, including federal and state natural resource agencies, the Port Liaison Project, local stakeholders, and nonprofit organizations, Port Orford Ocean Resource Team (POORT) has effectively taken several steps to implement a functional community based management structure in Port Orford: (1) education and empowerment of commercial fishermen to contribute to sustainable solutions, and (2) assessment of the biological, economic, and social health of the Port Orford reef, shoreline, and uses, including fisheries. As a result, fishermen, recreational users, and scientists have come together to plan and execute cooperative research and significantly expand the knowledge base of the Port Orford nearshore resources. The POORT and our partners recognize that participation by local ocean users in cooperative research can increase the scope and accuracy of stock assessments, reduce conflicts among various stakeholder groups, and build support for planning and implementing area management.

Josh A. Collins
Robert F. Heagey
Theodore S. Switzer
Florida Fish and Wildlife Conservation Commission
Fish and Wildlife Research Institute
100 8th Avenue SE
St. Petersburg, Florida 33701, USA

Abstract Title: Marine Reserves and Sport Fish: Collaboration between Scientists and Stakeholders
Authors: Collins, Josh; Heagey, Robert; Switzer, Theodore

The Florida Fish and Wildlife Conservation Commission's Fish and Wildlife Research Institute (FWRI) began a study in June 2004 to compare the distribution and abundance of inshore sport fish (primarily common snook *Centropomus undecimalis*, red drum *Sciaenops ocellatus*, and spotted seatrout *Cynoscion nebulosus*) in a de facto marine protected area and a nearby, unprotected area in Tampa Bay, Florida. To facilitate this research, FWRI scientists have collaborated with professional fishing guides and recreational anglers willing to volunteer both time and resources. Fishing guides frequently lead FWRI biologists on angling trips, providing researchers with a fisheries-dependent data set that can be used in comparing the size structure and abundance of sport fish within nearby protected and unprotected areas. FWRI biologists, along with two volunteer anglers, have tagged sport fish within both study areas to address important questions regarding fish movement and habitat use. Florida's recreational anglers have greatly augmented recapture rates; anglers reporting a recaptured fish receive a t-shirt and a personalized letter providing information about the project and the movements of the specific fish they recaptured. Our experiences indicate that such collaborative efforts enhance the overall quality and scope of research conducted while providing opportunities for direct communication between scientists and stakeholders.

Lisa L. Desfosse
National Marine Fisheries Service
1315 East West Highway, F/ST4
Silver Spring, Maryland 20910, USA

Abstract Title: National Cooperative Research Program: A Forum to Enhance National and Regional Cooperative Research Activities
Author: Desfosse, Lisa

The National Cooperative Research Program (NCRP) was initiated in 2001 by NOAA Fisheries Service's (NMFS) Office of Science and Technology. The program is supported by annual Congressional budget appropriations of approximately \$2,750K. Congressional appropriations are also provided to support cooperative research programs in the southeast, the northwest, and the northeast. The NCRP provides a coordination forum for cooperative research from all funding sources, as well as allocation of national funds. The majority of NCRP funds are used to support cooperative research activities conducted by the six NMFS fisheries science centers. Specific objectives of the NCRP include (1) provide a forum for national coordination of activities and use of results, (2) develop funding allocations for national cooperative research funds, (3) coordinate policy development, (4) enhance communication to showcase the successes of cooperative research, and (5) conduct national outreach activities. The NCRP has initiated the development of operational guidelines to improve the successful implementation of cooperative research activities. These guidelines will address the development of effective partnerships, the development of national and regional research priorities, planning and administration of projects, operational requirements, peer review of project results, dissemination of results, communication and outreach, and data management.

Abstract Title: Northeast Consortium: Building Partnerships and Delivering Data for Fisheries and Ocean Management
Authors: Gallant, Rachel

Rachel M. Gallant
Northeast Consortium
University of New Hampshire
142 Morse Hall
Durham, New Hampshire 03824, USA

The Northeast Consortium was created in 1999 to encourage and fund cooperative research and monitoring projects within the Gulf of Maine and Georges Bank that involve effective, equal partnerships among fishermen, scientists, and other stakeholders. The Northeast Consortium consists of four research institutions (University of New Hampshire, University of Maine, Massachusetts Institute of Technology, and Woods Hole Oceanographic Institution) that are working together to foster this initiative. Nearly $5 million from the National Oceanic and Atmospheric Administration is administered annually for cooperative research by the University of New Hampshire on behalf of the Northeast Consortium. Funds are distributed through an annual open competition, which is announced via a Request for Proposals. More than 150 cooperative research projects have now been funded, involving more than 280 fishermen and 100 scientists. Research topics include the design and testing of selective fishing gear, fisheries stock assessments, fish habitats, effectiveness of management measures such as closed areas and harvest limits, and the socioeconomic aspects of fishing. Priorities for funding change each year, as determined by the Advisory Committee's assessment of needs and opportunities for coastal, ocean, and fisheries management in the Gulf of Maine and Georges Bank. Commercial fishermen from Maine, New Hampshire, and Massachusetts participate. Commercial fishing vessels are being equipped and utilized as research platforms. The results from completed projects are impacting fisheries management decisions at local, state, and regional levels. The Northeast Consortium is working with the National Marine Fisheries Service and the New England Fisheries Management Council to ensure the use of new data, information, and products of cooperative research. Cooperative research has had a significant, positive impact of fishing communities and on the level of trust between fishermen, scientists, and fisheries managers; commercial fishermen are contributing to the scientific framework underlying fisheries management.

Abstract Title: Voices of the Fishing Industry: Communication, Culture, and the Interface with Fisheries Management
Authors: Gilden, Jennifer; Conway, Flaxen

Jennifer Gilden
Pacific Fisheries Management Council
7700 NE Ambassador Place, Suite 200
Portland, Oregon 97220, USA

Flaxen D. L. Conway
Oregon Sea Grant
Oregon State University
307 Fairbanks Hall
Corvallis, Oregon 97331, USA

This poster is based on two separate but related research projects. Semi-structured, ethnographic interviews were conducted for both projects. The first project examined communication and understanding between Oregon's fishing and fisheries management communities. The goal of this study was to evaluate communication strategies and tools used by both communities and to work toward improving communication in the future. The project identified existing communication strategies, determined how members of both communities perceive their effectiveness, explored how they affected the relationship between the communities, and identified and evaluated alternative communication strategies. Lack of communication both contributes to, and results from, the stress caused by fisheries crises. Products included *An Investment in Trust: Communication in the Commercial Fishing and Fisheries Management Community*, an Oregon Sea Grant publication. The second project explored obstacles to gathering and communicating about social and economic data and the reasons behind fishing community members' reluctance to participate in socioeconomic research. Products include a report that documents informal conversations with fishing community members about the importance of sharing socioeconomic data and discusses how surveys can be designed to improve response rates.

Pingguo He
University of New Hampshire
Institute for the Study of Earth, Oceans and Space
137 Morse Hall, 39 College Road
Durham, New Hampshire 03824, USA

Abstract Title: Cooperative Research at the University of New Hampshire: Working with Fishermen for Better Fishing Gears

Authors: He, Pingguo

Cooperative research in the northeastern United States has changed culture of fisheries in the region. Value sharing, experience recognition, and trust building between scientists and fishing industry participants has generated mutual understanding of fishing industry, fisheries science and fishery management. We take examples of cooperative research and training projects carried out between fishermen in the Gulf of Maine and scientists from University of New Hampshire to illustrate potentials of cooperation and long-term benefits to conservation and sustainable utilization of fisheries resources in the northeastern United States.

Scott A. Heppell
Department of Fisheries and Wildlife
Oregon State University
104 Nash Hall
Corvallis, Oregon 97330, USA

Ginny Goblirsch
Oregon Sea Grant
2030 SE Marine Science Drive
Newport, Oregon 97365, USA

Steven J. Parker
Oregon Department of Fish and Wildlife
2030 SE Marine Science Drive
Newport, Oregon 97365, USA

Kaety Hildenbrand
Marine Fisheries Extension
Lincoln County Extension Service
29 SE 2nd
Newport, Oregon 97365, USA

James W. Good
Marine Resource Management
Oregon State University
533 Weniger
Corvallis, Oregon 97331, USA

Abstract Title: Training Students to Be Effective Managers: Oregon State University's New Internship Program

Authors: Heppell, Scott; Goblirsch, Ginny; Parker, Steven; Hildenbrand, Kaety; Good, James

Oregon State University's Fisheries Training Internship is a unique course designed to provide undergraduate and graduate students with hands-on experience working with fishermen, processors, managers, researchers, and others who are intimately involved with the fishing industry and its management. The goal is to provide students with a working, insider's perspective on the industry and how it is supported and managed, from the factors that lead boat captains to decide where and when to fish to how resource management decisions are made. Initial opportunities included (1) at-sea commercial fishing experience, (2) at-sea fisheries research, (3) fisheries management experience, and (4) industry support facility experience. Key collaborators are proactive industry members who value training students in all aspects of fisheries management. In 2004, nine students enrolled in the first offering of this internship. All had unique, but equally valuable experiences and reported an in-depth understanding of the complex issues involved in marine fisheries management today. Financial and logistical support for this program is provided by the Department of Fisheries and Wildlife at Oregon State University, Oregon Sea Grant, Coastal Oregon Marine Experiment Station, The Oregon Department of Fish and Wildlife's Marine Resources Program, the Port Liaison Project, Cooperative Institute for Marine Resources Studies, and NOAA-Fisheries Northwest Fisheries Science Center/Fishery Resource Analysis and Monitoring Division.

Robert J. Hines
North Carolina Sea Grant
303 College Circle
Morehead City, North Carolina 28557, USA

Steve Rebach
North Carolina Sea Grant
North Carolina State University
1911 Building
Box 8605
Raleigh, North Carolina 27695, USA

Abstract Title: North Carolina's Fishery Resource Grant Program

Authors: Hines, Robert; Rebach, Steve

The Fishery Resource Grant Program was created in 1995 by the legislature to provide a mechanism for the state's fishing industries to participate in research concerning their means of livelihood and recreation. The program fostered a unique opportunity for commercial and recreational fishermen and aquaculturists to conduct research or partner with academic researchers. Four areas for research are allowed: fishing gear and equipment, environmental pilot programs, aquaculture, and seafood technology. Project selection is based upon merit, and geographic distribution as determined by an 11-member Grants Committee appointed from the state's Marine Fisheries Commission, its advisory committees, and the Sea Grant Program. The program has funded more than 300 projects, funding from 15 to 25 new projects each year. Results of completed projects have been used in fishery management plans for many fisheries and have assisted seafood businesses. The legislature provides 1 million dollars per year, 7.5% of which covers administrative costs. Spin-offs from the program have been the creation of the $500,000 Blue Crab Research Program and creation of a fishery resource grant program in Virginia.

Abstract Title: Community-Based Fisheries Management is Working in Maine—The Georges River Case Study
Author: Hoyt, Sherman

Sherman Hoyt
University of Maine Cooperative Extension/Sea Grant
377 Manktown Road
Waldoboro, Maine 04572, USA

From 1996 through 1999, the University of Maine Cooperative Extension and Sea Grant helped five Maine towns and their fishing communities develop a successful community-based fisheries management (CBFM) clam management program. University of Maine Cooperative Extension fisheries outreach staff provided capacity building assistance for the program. Since 1999, the program has been funded and run effectively by the communities themselves. As a result of this work, the towns and the state have legally granted the fishermen primary management responsibility—an essential element of CBFM and unique to the Georges River program. The program is self-supporting, using license fees and volunteer governance. It consistently produces the highest total landings of clams in Maine. The CBFM process protects water quality and clam populations very effectively. The participating communities are strengthened by the $3 million in commercial landings and jobs as well as the leadership capital and environmental stewardship ethic produced by the program. The fishery is sustainable and healthy. The University of Maine is working with and learning about CBFM from the Georges River case study. The success of the Georges River program offers compelling evidence that expanding CBFM into other fisheries could improve the effectiveness of fisheries management in Maine and New England.

Abstract Title: Using Collaborative Workshops to Improve the Management of Northern Shrimp in New England
Author: Hoyt, Sherman

Sherman Hoyt
University of Maine Cooperative Extension/Sea Grant
377 Manktown Road
Waldoboro, Maine 04572, USA

In April 2004, as part of the University of Maine Collaborative Fisheries Management Program, Cooperative Extension and Sea Grant staff from Maine, New Hampshire, and Massachusetts, in partnership with the Atlantic States Marine Fisheries Commission, conducted 3-d-long interactive workshops on improving shrimp management. Sixty fishermen, scientists, and managers from the three states participated. The workshops produced specific recommendations for improvements in the stock assessment process; more collaboration between the fishing industry, shrimp scientists, and managers; and the use of some new management tools for the New England shrimp fishery. Collaborative fisheries workshops do not follow the traditional stereotype of "experts" talking to attendees. The interactive format requires hard, thoughtful work from all participants. These workshops provide hands-on educational opportunities designed to maximize creative thinking and dialog with everyone actively engaged in resolving specific fisheries issues. The workshops become venues to increase public and fishing industry participation in fisheries management. We use small, neutrally facilitated groups to include participant ideas in a final report of the days work. The workshops provide new information and stimulate new ways of thinking about issues while avoiding the politicization often sought by participants to support their particular position on difficult fisheries issues.

Abstract Title: Cooperative Development of a Fisheries Management Plan in Biscayne National Park
Authors: Kellison, G; Clark, Rick; Tilmant, James; Robson, Mark

G. Todd Kellison
National Park Service - Biscayne National Park
Biscayne National Park
9700 SW 328th Street
Homestead, Florida 33033, USA

Clark Rick
Biscayne National Park
9700 SW 328th Street
Homestead, Florida 33033, USA

Tilmant, T. James
National Park Service - Water Resources Division
1201 Oak Ridge Drive, Suite 250
Fort Collins, Colorado 80525, USA

Located on the southeast coast of Florida, Biscayne National Park (BISC) encompasses 700 square kilometers, 95% of which is marine, and lies adjacent to metropolitan Miami, with a current and growing population of more than 2 million people. Multiple indices and stock assessments indicate that fisheries within BISC and in southeast Florida are in need of management and conservation actions. BISC is working cooperatively with the state of Florida's Fish and Wildlife Conservation Commission to develop a fishery management plan (FMP) for BISC that will benefit fishery resources within and beyond park boundaries. We discuss (1) the justification and need for cooperative management, (2) positive interactions and collaborations with other agencies (including the neigh-

Mark Robson
Florida Fish and Wildlife Conservation Commission
South Meridian Street
Tallahassee, Florida 32399, USA

boring Florida Keys National Marine Sanctuary), (3) the benefits of working cooperatively, and (4) the inclusion of public input in the decision-making process, including formation of a public advisory group consisting of commercial and recreational fishers, scientists, managers, and nongovernmental organizations representatives, among others. We suggest that the multiagency cooperative approach underlying FMP development and implementation provides an excellent protocol by which to develop strategies to responsibly manage and conserve fishery and ecosystem resources.

Abstract Title: A Cooperative Supplemental Transect Survey Targeting Mid-Atlantic Migratory Finfish Species
Authors: King, Sarah; Powell, Eric; Bochenek, Eleanor

Sarah E. King
Eric N. Powell
Eleanor A. Bochenek
Haskin Shellfish Research Laboratory
Rutgers University
6959 Miller Avenue
Port Norris, New Jersey 08349, USA

One goal is to improve stock abundance estimates by augmenting survey data that will enhance the underlying database supporting abundance estimates. The main objective of this study is to supplement the National Marine Fisheries Service (NMFS)-Northeast Fisheries Science Center (NEFSC) survey database with information on the deepwater extent and migratory behavior of recreationally and commercially important species. A mid-Atlantic multispecies survey program was developed that extends beyond the domain of present-day federal and state surveys with increased sampling intensity between 150 and 250 fathom using a commercial fishing vessel. The survey is a cooperative effort among the commercial fishing industry, National Fisheries Institute-Scientific Monitoring Committee, Rutgers University, and NMFS-NEFSC. To date, there have been seven sampling events on two transects sited near Baltimore and Hudson canyons. The NMFS assists with fieldwork and program design, the industry provides some financial and logistical support, and Rutgers University oversees the project and conducts fieldwork, data entry, and analysis. Data collected during these cooperative surveys are audited, summarized in a cruise report, and sent directly to the NMFS-NEFSC survey database by Rutgers University. Data are used by the Mid-Atlantic Fishery Management Council and NMFS to help improve abundance estimates of target species, which include Loligo squid *Loligo* spp., spiny dogfish *Squalus acanthias*, goosefish (also known as monkfish) *Lophius americanus*, summer flounder *Paralichthys dentatus*, silver hake *Merluccius bilinearis*, offshore hake *M. albidus*, scup *Stenotomus chrysops*, and black sea bass *Centropristis striata.*

Abstract Title: Archival Tagging of North Pacific Albacore: The Latest Success in More than 30 Years of Cooperation with the U.S. Albacore Fishery
Authors: Kohin, Suzanne*; Childers, John; Sakagawa, Gary

Suzanne Kohin
John Childers
Gary Sakagawa
NOAA Fisheries, Southwest Fisheries Science Center
8604 La Jolla Shores Drive
La Jolla, California 92037, USA

Since 1971, the Southwest Fisheries Science Center (SWFSC) has had an ongoing partnership with the West Coast-based U.S. albacore *Thunnus alalunga* fishing industry. Research is conducted in cooperation with the American Fisherman's Research Foundation (AFRF), a private foundation established by the Western Fishboat Owner's Association to promote research on albacore and related fish. Past projects have included exploratory fishing with simultaneous collection of oceanographic data, development of a standard logbook, research into the impact of high seas driftnetting, onboard and port sampling for size frequencies, and conventional tagging to study movements, growth, and longevity. Since 2001, SWFSC and AFRF have been conducting an archival tagging project to study migratory patterns, depth, and temperature preferences of north Pacific albacore. To date, 278 tags have been deployed with the goal of deploying 500 tags by 2006. Sixteen tags have been recovered, and the data demonstrate extensive migratory movements and diurnal vertical excursions to beyond 200 m during daytime with fish remaining in the upper 50 m at night. SWFSC's cooperative relationship with AFRF and the albacore fishing fleet is effective because the results contribute to common goals of both groups: understanding more about the biology of albacore and the status of the stocks.

Abstract Title: Fisheries Priorities and Cooperative Research: Working for a Common Goal
Author: La Valley, Ken

Ken La Valley
University of New Hampshire Sea Grant
Nesmith Hall
Durham, New Hampshire 03824, USA

Since 2000, more than 20 million dollars have been awarded to more than 170 cooperative research projects in the northeast. The majority of cooperative research projects have been supported by two regional programs, the Northeast Consortium and the Cooperative Research Partnership Initiative (CRPI). The Northeast Consortium, which was created in 1999, promotes equal partnerships among commercial fishermen and scientists and focuses research in the Gulf of Maine and Georges Bank. The Northeast Consortium receives its funds from the National Oceanic and Atmospheric Administration. The Northeast Regional Office of the National Marine Fisheries Service developed the CRPI in 1999 to support and enhance collaborative research among the New England commercial fishing industry, scientists, and management communities. The Northeast Consortium funds a broader scope of projects that range from oceanographic monitoring to fishing gear conservation engineering, while the initiative of the CRPI is to enhance the data upon which fishery management decisions are made. A GAP analysis was conducted, which compared the published fisheries priorities identified by fishing, scientific, conservation, and management communities over the past 4 years through public scoping meetings and other public stakeholder discussions, to the portfolio of awarded cooperative fisheries research projects.

Abstract Title: Scientists and Fishermen Exchange
Authors: McMullen, Scott; Hildenbrand, Kaety*; Goblirsch, Ginny; Conway, Flaxen; Harms, John

Scott McMullen
Oregon Fishermen's Cable Committee
2021 Marine Drive, Suite 102
Astoria, Oregon 97103, USA

Kaety Hildenbrand
Ginny Goblirsch
Oregon Sea Grant
29 SE 2nd
Newport, Oregon 97365, USA

Flaxen D. L. Conway
Oregon Sea Grant
Oregon State University
307 Fairbanks Hall
Corvallis, Oregon 97331, USA

John Harms
NOAA Fisheries, Northwest Fisheries Science Center
2725 Montlake Boulevard East
Seattle, Washington 98112, USA

Distrust is the biggest challenge to relations between the fishing and science communities. Uncertainty and distrust in the fishing community, and overwork, bureaucracy, and a lack of time and funding in the science/management community, lead to cynicism and poor involvement and cooperation with research and regulations. A commonly expressed perception that erodes trust is that managers and scientists really are not interested in listening to the fishing community. Both communities have expressed doubts about available knowledge and its interpretation and use. Oregon Sea Grant designed and developed an innovative venue for improving science and relationships between today's—and tomorrow's—commercial fishermen and ocean scientists. The Scientists and Fishermen Exchange (SAFE) is a comfortable venue where genuine discussion and information exchange occurs between fishermen and scientists, ideas are developed, relationships are built/strengthened, and a mutual understanding of the industry and ocean sciences occurs. This kind of knowledge exchange and interaction between fishermen and marine researchers is crucial. The goal with SAFE is to create an atmosphere so comfortable that, over time, open/frank/respectful exchange can take place on even the most contentious issues. SAFE is a place to listen, learn, and understand diverse viewpoints while getting to know, respect, and understand each other.

Abstract Title: Collaborative Systems Approach to Prioritizing Fisheries Research and Harvest Monitoring
Authors: Merritt, Margaret; McBride, Doug; Fried, Stephen; Wheeler, Polly; Craver, Amy

Margaret F. Merritt
University of Alaska Fairbanks
882 Vide Drive
Fairbanks, Alaska 99712, USA

Doug McBride
Stephen Fried
Polly Wheeler
Amy Craver
USFWS Office of Subsistence Management
3601 C Street, Suite 1030
Anchorage Alaska 99503, USA

To ensure strategic use of Federal Subsistence Fisheries Monitoring Program funds, the U.S. Fish and Wildlife Service's Office of Subsistence Management initiated a collaborative process to identify and prioritize program goals, research objectives, and information needs, using a systems approach to strategic planning with decision support software. Facilitated workshops were convened in 2004–2005 for the Copper River–Prince William Sound and the Bristol Bay–Chignik areas of Alaska. Participants included social and biological scientists from government and academia, anthropologists, fisheries managers, members of regional native organizations, and regional advisory council members. While a systems approach to strategic planning has been used extensively in such fields as business and military and medical sciences, applications to fisheries problems are recent. Benefits from using a systems approach to strategic planning include an easily understood and communicable framework of ideas, transparency in expert judgments and an increased likelihood for an optimal solution. Several factors contributing to successful collaboration of stakeholders in the decision-making process are presented.

Scot B. Plank
State/Federal Liaison Office
National Marine Fisheries Service
263 13th Avenue South
St. Petersburg, Florida 33701, USA

Ellie F. Roche
National Marine Fisheries Service
Koger Building
9721 Executive Center Drive
North
St. Petersburg, Florida 33702, USA

Abstract Title: Caribbean Fishing Recommendations Developed as a Result of Funding by the Southeast Cooperative Research Program

Authors: Plank, Scot; Roche, Ellie

The Southeast Cooperative Research Program is sponsored by the National Marine Fisheries Service Southeast Fisheries Science Center in Miami and assisted by the State/Federal Liaison Branch in St. Petersburg, Florida. The intent of the cooperative research program in the southeast region is to utilize the collective experience of fishermen and scientists to produce the best advice to fishery managers based on fishing experience and sound scientific research procedures. Interaction between fishermen and scientists benefits the concept and implementation of research studies and also improves the public acceptance of scientific results that are produced. A case in point is the Fiscal Year 2003 project competitively awarded to MRAG Americas, Inc. During the course of the project, federal funds were spent to complete workshops to determine fishers' attitudes toward potential effort reduction programs in the Caribbean region of the United States. Working together, fishermen and scientists were able to evaluate fishery resources and practices and prepare recommendations for fishery managers as the end result. After completion of the final report in September 2004, MRAG Americas, Inc. provided the project recommendations to the Caribbean Fishery Management Council.

Scot B. Plank
State/Federal Liaison Office
National Marine Fisheries Service
263 13th Avenue South
St. Petersburg, Florida 33701, USA

Ellie F. Roche
National Marine Fisheries Service
Koger Building
9721 Executive Center Drive
North
St. Petersburg, Florida 33702, USA

Abstract Title: Research Applications through the Southeast Saltonstall-Kennedy Program

Authors: Plank, Scot; Roche, Ellie

The Saltonstall-Kennedy Program in the southeast region has competitively funded cooperative agreements with biologists in academia and state agencies, fisheries organizations, and industry partners. Selected projects have supported research in the commercial fishing industry, methods for safer and wholesome seafood, and sustainable fisheries. The author will present case studies of several successful cooperative agreements that demonstrate the success of the program.

Robert A. Robertson
Troy W. Hartley
Department of Resource Economics and Development
University of New Hampshire
56 College Road
Durham, New Hampshire 03824, USA

Abstract Title: Cooperative Research and the Management of Marine Fishery Resources in New England: How Important Is Trust?

Authors: Robertson, Robert; Hartley, Troy

Cooperative research has been held up by some advocates from the fishing, scientific, management, and policy-making communities as a way to incorporate the knowledge of commercial fishermen into the scientific assessments of marine fisheries. Cooperative research seeks to promote partnerships between fishermen and researchers to improve the quality of data used in fisheries management and to improve the relations between the fishing industry and the scientific/management communities. The success of cooperative research initiatives in achieving these goals depends, at least in part, on the degree to which individuals trust each other at an individual or group/agency level. There are many potential barriers to cooperative research and very little empirical research on the socioeconomic or human dimensions of this topic. This paper will provide a preliminary understanding of the challenges and prospects of cooperative research from the perspective of members of the commercial fishing, scientific, and management community.

Abstract Title: An Electronic Logbook Developed through Cooperative Research to Support Shrimp Fishing and Fisheries Management

Authors: Rubec, Peter; Ashbaugh, Charles; Versaggi, Salvatore; Weisberg, Robert; Lewis, Jesse; Reed, David; Jenkins, Chris

Peter J. Rubec
Jesse M. Lewis
David Reed
Florida Fish & Wildlife Conservation Commission
100 Eighth Avenue SE
St. Petersburg, Florida 33701, USA

Charles F. K. Ashbaugh
Sasco Inc.
4101C East 12 Avenue, Suite 2
Tampa, Florida 33605, USA

Salvatore Versaggi
Versaggi Shrimp Corporation
2633 Causeway Boulevard
Tampa, Florida 33619, USA

Robert H. Weisberg
University of South Florida, College of Marine Science
140 7th Avenue South
St. Petersburg, Florida 33701, USA

Chris Jenkins
University of Colorado at Boulder
Institute for Arctic & Alpine Research
1560 30th Street
Boulder, Colorado 80309, USA

An electronic logbook (ELB) was developed by Florida Fish and Wildlife in cooperation with Versaggi Shrimp Corp. and Sasco Inc. The ELB was used by the captain to record catch, effort, and environmental data onto a computer situated in the wheelhouse of the boat. A data logger (CTD) associated with the boat's trynet recorded vertical and horizontal profiles. The CTD data were transferred wirelessly to the computer each time the trynet was retrieved. Counts of shrimp numbers from the trynet and weights of size grades of four shrimp species obtained from the main trawls were recorded using the ELB system. Positions (latitude and longitude) automatically recorded by the boat's GPS receiver allowed determination of fishing effort (hours fished). Environmental data collected by the CTD were displayed on screen to depict changes in temperature or salinity over the path taken by the trynet. Oceanographic data collected by the University of South Florida tied to the Coastal Ocean Monitoring Prediction System were modeled and mapped for 16 seasons extending from March 2004 to the end of June 2005. The data were analyzed to conduct habitat suitability modeling (HSM) of the pink shrimp *Farfantepenaeus duorarum* fishery on the west Florida shelf. The ELB system maintains a database of use to the captain and provides data to support HSM analyses and geographical information system-based mapping. The ELB system is useful to the fishermen and provides catch, effort, and habitat-related information of potential use for fisheries managers to better manage the shrimp fishery and essential fish habitat.

Abstract Title: Developing Practical Knowledge through Collaborative Research—A Coast-Wide Study in California and Oregon

Authors: Schlosser, Susan; Bloeser, Jennifer

Susan C. Schlosser
University of California Sea Grant Program
2 Commercial Street, Suite 4
Eureka, California 95501, USA

Jennifer A. Bloeser
Pacific Marine Conservation Council
Post Office Box 327
Arcata, California 95521, USA

State and federal legislation requires fishery management plans to include life history information of managed species. We recognized the lack of data on juvenile nearshore rockfish and cabezon *Scorpaenichthys marmoratus*. Working with local fishermen in Eureka, California, we designed a collaborative sampling design to study nearshore habitats not accessible by trawl. The pilot project in Humboldt Bay was conducted for 1 year. The pilot project was successful. We received funding for a coast-wide project to expand the project to nine ports between Newport, Oregon and Morro Bay, California. Fishermen, numerous state and federal agency biologists, and Sea Grant marine advisors are among project collaborators. Fishermen are paid for monthly sampling. Through the Port Liaison Project, four fishermen have been trained to conduct monthly sampling. Knowledge was increased on the use of nearshore habitats by early benthic juvenile rockfish and cabezon. Mutual education of project collaborators continues to occur. Annual reports are disseminated to all project collaborators and many who follow the project. Telephone calls and quarterly conference calls are the primary communication among project collaborators. We are completing our third year of sampling and plan to identify index sites for to establish permanent sites of juvenile rockfish and cabezon sampling in 2005.

Abstract Title: Lessons Learned: Perspectives on Facilitating Collaboration among Fishermen and Scientist in the Gulf of Maine

Authors: Singer, Laura

Laura Taylor Singer
Gulf of Maine Research Institute
350 Commercial Street
Portland, Maine 04101, USA

The 21st century has brought with it new and revitalized efforts to conduct fisheries research that involves scientists and fishermen as research partners. Collaborative research (or cooperative research) is not a new concept, but it has taken hold in New England due to increased funding opportunities. Other areas of the country are looking for ways to increase their use of this approach.

Successful collaborative research requires special skills in engaging the scientific and fishing community in a new dialog of shared learning. This poster will address how to work through the challenges of developing and implementing a successful collaborative research partnership, highlight the creative possibilities these new partnerships can bring to fisheries research, and review the key elements of some recent effective collaborative research partnerships in New England.

Laura G. Skrobe
Kathleen Castro
David Beutel
Barbara Somers
Rhode Island Sea Grant
University of Rhode Island
Fisheries Center East Farm
Building 83
Kingston, Rhode Island 02881, USA

Abstract Title: Finding Solutions Together: Combining Outreach and Research to Resolve Bycatch Issues in the Southern New England Fisheries

Authors: Skrobe, Laura; Castro, Kathleen; Beutel, David; Somers, Barbara

One of the key missions of the Rhode Island Sea Grant (RISG) Sustainable Fisheries Extension Program is to engage and enable stakeholders to play critical roles in science and management of the fisheries. This is accomplished through research, education, and outreach. The research component involves conducting applied collaborative research on issues that elucidate processes and link solutions with the effective science and management of the resource and the fisheries. Projects such as trawl cod end mesh selectivity studies on yellowtail flounder *Limanda proboscidea*, winter flounder *Pseudopleuronectes americanus*, and summer flounder *Paralichthys dentatus*, bycatch characterization study on scup *Stenotomus chrysops*, and pot selectivity study on increasing escape vent size on black sea bass *Centropristis striata* and scup are good examples. Outreach projects such as the Regional Bycatch Workshop and Gear Conservation Engineering Network provide stakeholders the ability to input their knowledge into the process as well as receive information from various other organizations on fisheries science and management. The RISG fisheries program has a long successful history contributing to the solutions for bycatch problems in Rhode Island and the New England region through collaborative research and outreach with fishermen, environmentalists, and the management agencies. This poster will address major results of these research and outreach projects.

Kevin D.E. Stokesbury
Bradley P. Harris
Michael C. Marino II
Jacob I. Nogueira
Christopher L. Sarro
Michelle E. Schenk
Melissa A. Allard
Brent M. Courchene
Thomas A. Jaffarian
Christel Flis
University of Massachusetts
Graduate School for Marine Science and Technology
706 South Rodney French Boulevard
Dartmouth, Massachusetts 02744, USA

Abstract Title: Conservation through Cooperation: 7 Years of UMass—Commercial Sea Scallop Industry Cooperative Research

Authors: Stokesbury, Kevin; Harris, Bradley; Marino II, Michael; Nogueira, Jacob; Sarro, Christopher; Schenk, Michelle; Allard, Melissa; Courchene, Brent; Jaffarian, Thomas; Flis, Christel

Since 1999, the University of Massachusetts School for Marine Science and Technology and the commercial sea scallop industry completed 48 video and 5 tagging cruises on Georges Bank and in the mid-Atlantic. We examined scallop populations and benthic habitat in 81,000 video samples covering 60,000 km^2 of the northeast U.S. continental shelf and tagged 30,000 scallops to assess shell growth. Further, we dissected 10,000 scallops to measure temporal and spatial variability in shell height to meat weight relationships. Scallop industry members helped develop sampling methods and equipment, participated in video survey and tagging cruises, and provided live scallops for dissection. The commercial scallop industry donated more than 6,000 h of vessel time, 2,000 meals, 50,000 gallons of fuel, and US$120,000 to pay captains and crew participating in surveys. Seventy-four different commercial fishers have participated in weeklong research cruises. The Massachusetts Division of Marine Fisheries and NOAA Fisheries provided additional support. The data from this research were used extensively in sea scallop *Placopecten magellanicus* fisheries management and habitat assessment, and the 1999 and 2003 scallop survey data were published in peer-reviewed scientific journals.

Abstract Title: The Northeast Regional Cod Tagging Program: A Strong, International Partnership of Scientists, Fishermen, and Managers
Author: Tallack, Shelly

Shelly M. L. Tallack
Gulf of Maine Research Institute
350 Commercial Street
Portland, Maine 04102, USA

The Northeast Regional Cod Tagging Program (NRCTP) represents a significant example of international collaborative research, involving ~250 fishermen and ~25 scientists from Nova Scotia to Cape Cod. This 3-year program on Atlantic cod *Gadus morhua* is the largest tagging study initiated along the eastern seaboard of North America. Based on scoping study recommendations documented by the New England Aquarium, the NRCTP aims to (1) improve understanding of cod movements in the Gulf of Maine, Georges Bank, and neighboring New England and Canadian waters; (2) tag 100,000 cod throughout the region using a standard tagging technique; (3) obtain contemporary growth information; (4) encourage the participation of commercial and recreational stakeholders; (5) build an online database with accessibility through a geographical information system mapping interface (www.codresearch.org); and (6) identify and develop testable hypotheses for further research. The program has released more than100,000 tagged cod throughout the study region between March 2003 and December 2004. Collaboration has been high, with more than 75 commercial and recreational fishing vessels working in association with six tagging organizations. Tag returns have steadily increased, yielding valuable information on cod movement and growth. These data will also assist the New England Fishery Management Council to manage the multispecies groundfish complex.

Abstract Title: Increasing the Awareness and Understanding of Highly Migratory Species Federal Regulations for Recreational Anglers
Authors: Thigpen, Jack; Smutko, Steve

Jack Thigpen
North Carolina Sea Grant
North Carolina State University
Box 8605
Raleigh, North Carolina 27695, USA

Steve Smutko
North Carolina Sea Grant
North Carolina State University
326 Nelson Hall
Box 8109
Raleigh, North Carolina 27695, USA

North Carolina Sea Grant collaborated with the Highly Migratory Species Management Division of NOAA Fisheries on a pilot project to better understand recreational anglers information needs regarding highly migratory species in Atlantic and Gulf of Mexico waters. A professionally facilitated workshop was conducted North Carolina and brought together a purposively selected group of opinion leaders from the following communities: offshore charter boat captains, private offshore recreational anglers, offshore fishing tournament organizers, Sea Grant fisheries extension specialists, National Marine Fisheries Service (NMFS) fisheries managers and fisheries management staff from the North Carolina Division of Marine Fisheries. They were led through a facilitated process to determine the best ways for NMFS to reach recreational anglers with information on highly migratory species and identify the communications issues related to new regulations affecting theses species. The primary focus was on finding ways to improve the collection and dissemination of information on recreational fisheries management for bluefin tuna *Thunnus thynnus* and billfishes. Quotas, permitting, gear requirements, and data collection and reporting were discussed at length. The panel completed its task by developing a set of 11 recommendations for improving NMFS communications and 2 recommendations related to improvements in management and enforcement.

Abstract Title: Making the Fisheries Extension Connection: Sea Grant Extension and the U.S. Commercial Shrimp Industry Crisis
Author: Von Harten, Amber

Amber Von Harten
South Carolina Sea Grant Consortium
Post Office Box 189
Beaufort, South Carolina 29901, USA

The U.S. commercial shrimp industry is undergoing an economic crisis fueled by changing market conditions and the influx of imported shrimp from foreign countries. In response to the crisis, the commercial shrimp industry in South Carolina was awarded $5.6 million in economic assistance by the federal government in 2003. The use of the awarded funds was determined utilizing input from an industry-derived task force (South Carolina Shrimp Industry Task Force) along with advisors from state and private organizations and facilitated by Clemson Extension Service. Coincidentally, the South Carolina Sea Grant Consortium (SCSGC) reinstated the Fisheries Extension and Enhancement (FEE)

Program in 2003 to respond to the state's growing extension needs related to marine fisheries. The FEE program at the SCSGC has been instrumental in facilitating the communication and exchange of industry-related research, state/federal fisheries management information, and technical assistance to the commercial shrimp industry in South Carolina. Currently, the FEE program is identifying additional extension needs in the recreational/commercial fishing sectors in the state to include collaboration with state and federal fisheries management agencies. Results from the efforts of the SCSGC-FEE program will allow targeted commercial and recreational stakeholders to make the best informed decisions regarding marine resources using science-based information.

Part IV: Concluding Remarks

American Fisheries Society Symposium 52:251–255, 2006

The Future of Partnerships for a Common Purpose: Cooperative Fisheries Research and Management

Troy W. Hartley
Northeast Consortium
Department of Resource Economics and Development
University of New Hampshire
310 James Hall, Durham, New Hampshire 03824, USA

Alesia N. Read
American Fisheries Society
5410 Grosvenor Lane, Suite 110, Bethesda, Maryland 20814, USA

The *Partnerships for a Common Purpose: Cooperative Fisheries Research and Management* symposium was one step in a longer process of reviewing, analyzing, and discussing the rapidly evolving and expanding phenomenon of cooperative fisheries research and management. Fisheries are at a crossroads in America and worldwide. Calls for change have come from major national reviews (e.g., Pew Commission and U.S. Ocean Commission), and there exists a broader recognition among industry, scientists, managers, and other fisheries stakeholders that, in spite of progress on some fronts, we can do better.

The models and tools of comanagement have expanded and today include multinational commissions, maturing regional councils, new sector and area management initiatives, and community-based management programs. Most of these cooperative initiatives have emerged from tremendous conflict and contention. Change is not easy, and there will be more tension in the future of fisheries research and management. Further, the explosion of cooperative fisheries research throughout the United States and Canada, but particularly profuse in the northeast Atlantic, the Pacific Northwest, and Alaska, might be attributable, in part, to the discord and dysfunction of federal fisheries management in the 1980s and into the 1990s. However, today, there is a growing expectation that for us to do better, for us to more fully understand the ocean and marine environment and successfully study and manage marine ecosystems, we must work together. No single stakeholder can do it alone. Fishermen, scientists, managers, and other stakeholders must find ways to integrate what they know, design and conduct research together, and manage the resources together for the betterment of all stakeholders, the general public, and our children.

While the *Partnerships for a Common Purpose: Cooperative Fisheries Research and Management* symposium took an important step forward, much is left to discuss, study, and learn. Specific impacts were identified; many reflect hypotheses worthy of future research. Many lessons learned have emerged from the experience of others, and these form the basis for a general how-to guide of principles that acknowledge that a one-size-fits-all model will not work since local, state, and regional socioeconomic, political, cultural, and ecological context will determine the most appropriate details of collaboration. A wide range of future research and policy questions were identified throughout the discussions. The symposium participants acknowledged the harsh realities of cooperation, the challenges, and the lessons learned the hard way. The road ahead will

“No single stakeholder can do it alone. Fishermen, scientists, managers, and other stakeholders must find ways to integrate what they know, design and conduct research together, and manage the resources together for the betterment of all stakeholders, the general public, and our children.”

have many bumps. But throughout it all, participants were optimistic about the potential for cooperative fisheries research and management and about the future of fish, fishermen, knowledge, and our ability to manage the resource and the ecosystem in a sustainable manner for future generations. We conclude by summarizing what symposium participants told us about the impacts, lessons learned, and future research and policy issues and questions central to moving forward and integrating cooperative fisheries research and management.

Impacts

Participants in cases of cooperative fisheries research and management and those attending the symposium clearly identified specific human dimension impacts, along with broader, more general scientific and management achievements (see Table 1). There was a general acknowledgement among symposium participants that while there must be a commitment to rigorous science, which can be time consuming, more needs to be done to quickly integrate cooperative research findings into management actions. A few examples existed of specific research-to-action impacts, such as the increased protection of sea turtles through turtle excluder devises and the direct industry funding of protection for Kemp's ridley nesting areas, which have improved species abundance. The Nordmore grate in the shrimp industry has dramatically reduced bycatch and discards.

Nonetheless, the future looks bright, as Jim Murray with the National Sea Grant Office said, "Often change is incremental in society and government, but progress is being made." Jim Balsinger, acting deputy assistant administrator for regulatory programs for NOAA Fisheries, added,

> "Ecosystem based management has been helped by cooperative research, because we [now] engage different stakeholder groups in the process and the questions are much broader. You cannot just look at uses of the natural marine resources in the ocean. You have to also look at the processing industry, the communities, the shipping industry. These things may not have been brought to the table without cooperative research. It is a start. Ecosystem based management is going to extend beyond the traditional way of management."

Lessons Learned and the How-To's of Cooperation

Cooperative fisheries research and management are not easy. They take considerable time, effort, coordination, and follow-through. The list of challenges can be long and disheartening: time; funding; adequate motivation; sufficient and capable participants; adequate communication; information management; and many other obstacles specific to individual projects. Nonetheless, the symposium discussions also identified tips and strategies for overcoming challenges, summarized below in Table 2. These form the basis for a general how-to guide of principles that acknowledge that a one-size-fits-all model will not work since local, state, and regional socioeconomic, political, cultural, and ecological context will determine the most appropriate details of collaboration.

The Future

Table 3 summarizes the future research and policy issues and questions identified at the symposium. It is not an all-inclusive list of needs and challenges ahead, but rather a subset of issues and questions that rose to the surface among the more than 150 participants in the case studies and symposium sessions. These are topics for future workshops, conference symposia, research projects, and management discussions among all the interested parties, fishermen, scientists, managers, nongovernmental organizations, and citizens.

Finally, as opportunities arise to conduct comanagement experiments on smaller scales in the United States (e.g., area and sector management), a commitment is needed to conduct science on the same scale and build the real-time feedback loops into the research and management systems.

Table 1. What did cooperative fisheries research and management accomplish?

Management impacts and achievements—particularly from cooperative management

- Produces *tangible outcomes*, solving problems, expanding knowledge, and producing efficiencies, including the following:
    - Creates more expansive, holistic management view of the problem.
    - Generates broad data sets that provide ability to employ and test adaptive management strategies.
    - Builds network for ecosystem-based management.
    - Produces greater range of management options, including more shared responsibility among parties, and additional tools and practices to reduce bycatch and fish mortality.
    - Results in better understanding of fishermen perspectives and actual fishing practices.
    - Leads to productive, healthy fish stocks and protected and restored ecosystem health.
- Improves *decision-making climate* among fishermen, scientists, and managers.
    - Improves decision-making climate, less contentious dialog.
- Effective problem-solving with consensus agreements and less litigation.
    - Reduces need for enforcement, promoting greater compliance and greater support for actions.
    - Enhanced coordination and consolidation of management actions.

Scientific impacts and achievements—particularly from cooperative research

- Produces *tangible outcomes,* understanding, capacity, and stock assessments.
    - Achieves specific scientific research goals, including answers to specific basic and applied research questions and long-term data sets.
    - Enhances research capacity, financial efficiency, coordination.
    - Enhances understanding of science and fishermen's knowledge.
    - Increases interdisciplinary research and expands scope of research questions.
    - Expands knowledge base, including greater scientific knowledge and inclusion of local knowledge.
    - Increases understanding of stock structure, including additional stock assessments and bycatch and fish mortality levels.
- Improves *scientific administration and output.*
    - Enhances coordination and consolidation of research.
    - Provides greater support for scientific findings.

Human impacts and achievements

- Expands *industry involvement,* particularly in scientific research.
    - Expands industry capacity and involvement.
    - Promotes broader cooperation and understanding between industry sectors.
    - Supports alternative income sources for fishing communities.
    - Provides venue for additional stakeholder involvement (e.g., conservation organizations, recreational fishing sector).
    - Empowers fishermen.
- Builds *trust and respect* between fishermen, scientists, and managers.
    - Improves mutual understanding, mutual respect, and trust.
    - Builds common language.
    - Builds partnership capacity.
- Facilitates *broader ecosystem perspectives* and *shared responsibility* for the future.
    - Promotes a sense of stewardship among resource users.
    - Shares and integrates scientific and fishing industry knowledge.
    - Promotes new ideas, creativity.
    - Provides opportunities to bridge diverse professional cultures.
- Generates *hope, optimism, and excitement* for the future.

Consequently, exceptional case studies of integrating cooperative fisheries research and management may emerge in the design and implementation of area and sector management pilot projects currently underway throughout the United States. The lessons from these comanagement experi-

Table 2. How-to's of cooperative fisheries research and management.

Getting cooperative fisheries research of management started: to date, new programmatic opportunities and enhanced funding for partnerships have made it easier to start cooperative research. Broader reforms in management regimes and institutions have kicked off comanagement initiatives. Once such institutional change has occurred and there are opportunities for collaboration, there are some how-to tips to consider:

- *Extensive planning*—invest the substantial time into planning to avoid future pitfalls.
    - Develop *clear, relevant, and agreed-upon objectives.*
    - Identify *distinct roles and responsibilities* for all partners to ensure that expectations are realistic and understood by all.
    - *Engage partners* early and often, and if necessary, expand stakeholder group and participant engagement during implementation. Some partners may need more assistance because their organizational infrastructure and capacity is less developed. Provide wide range of options for participating (i.e., multiple role and responsibilities available for prospective participants).
- *Address partners' needs and interests*—multiple incentives are needed to ensure that cooperation is sustained, such as the following interests raised by symposium participants:
    - Leverage resources and promote *efficiencies*, including meeting funding and employment needs.
    - Enhance *likelihood* of change and genuine impact.
    - Respond to new *expectations* in today's fisheries research and management.
    - Produce greater synergistic benefits and *multidisciplinary capacity*, including engaging industry and integrating fishermen and scientific knowledge.
    - Build *mutual understanding and trust.*
    - Promote an individual's *alternative* viewpoint or perceived need for an integrated management strategy, including concerns with existing knowledge or management system.
    - *Learn* more about the resource, particularly among cooperative research participants.
    - Advance an individual's *sense of duty and responsibility.*

Managing operations: while individual examples of cooperative research and comanagement varied tremendously, there were a few common implementation features that helped navigate the challenges inherent in collaboration.

- Maintain *flexible approaches* to achieve clear, distinct goals.
- Ensure adequate *time management and planning*, often requiring more time than expected.
- Ensure adequate *staff capacity* to manage people, information, and communication.
    - Employ project coordinators, and outreach, education, and data management staff.
    - Ensure adequate leadership that can deal with change and uncertainty.
    - Employ sufficient coordinating mechanisms.
- Generate and apply clear *decision-making protocols* and operating procedures.
- Employ multiple communication tools and mechanisms to advance *open, transparent, and inclusive communication.*
- Promote constructive professional *working relationships.*
    - Maintain clear roles and responsibilities.
    - Engage skilled and capable partners.
    - Attend to partners' needs and interests so that they remain motivated and engaged.
    - Celebrate success.
- Be creative to leverage and obtain adequate and *consistent funding.*
- Maintain high *professional quality standards.*
    - Monitor progress.
    - Adhere to the latest safety standards, insurance, contracts, compensation standards, and practices.
    - Conduct independent, peer review, audits, etc.

Avoiding pitfalls during implementation: many of the examples of cooperative research and comanagement discussed in the symposium were considered successful by the participants. They attended to the getting started and implementation tips above. Nonetheless, they still had to weather challenges and often did so successfully. Below are common challenges faced in cooperative fisheries research and comanagement:

Table 2. Continued.

- Inconsistent funding which *undermines commitment* to partnership.
- *Integrating fishermen's knowledge* with the scientific or management framework.
    - Obtaining acceptance and buy-in from affected parties.
    - Finding feasible solutions that meet party needs.
    - Overcoming divergent interests.
- Reconciling *data sharing and data control*, particularly in cooperative research.
- Adequate *management, communication, and coordination.*
    - Managing the volume of information and data.
    - Coordinating all the parties.
    - Securing adequate time to fully participate.
    - Securing and engaging capable partners.

ments may contribute significantly to facilitating integration of cooperative fisheries research and management on broader regional scales.

The future is now for cooperative fisheries research and management. They require communication among all the stakeholder groups, including a lot of listening and questions from all parties, but the rich dialog that occurred during the *Partnerships for a Common Purpose: Cooperative Fisheries Research and Management* symposium bodes well for the future.

Table 3. Future research and policy questions and issues for cooperative fisheries research and management.

Ecosystem-based management

- How to build real-time *feedback loops* to fishermen, scientists, and managers that enable adaptive management, ongoing monitoring, and adjustments in research priorities?
- How to build nested *governance structures* that enable experimental cooperative fisheries management pilots on a finer geographic, sector, and/or ecological scale?
- How to *protect the rights* of less powerful and prevent inequities and injustice?
- What are the *lessons learned* from the current pilot studies in area and sector management underway throughout the United States? How is nested cooperative management structured? How is research coordinated? How are priorities balanced? How are real-time feedback loops designed and implemented, and how well do they work?

Human dimensions

- How is *trust* built? What undermines trust and how?
- What makes research findings, the scientific framework, and management decisions and system *legitimate and credible* in the eyes of each party?
- What role is *fairness* playing in trust, legitimacy, credibility, and effective cooperative fisheries research and management?
- Who *participates* and how? Whose responsibility is it to ensure broad participation? What variety of participation options are needed? Who is permitted to fish and who might not be? How is account ability for the resource shared in a cooperative fisheries research and management system?
- How to *integrate knowledge* from multiple disciplines and with fishermen and other types of local ecological knowledge?
- What role is *communication* playing in partnership development, trust building, and the integration of knowledge? How to effectively communicate between diverse knowledge structures? How does transparency in communication contribute to trust, legitimacy, credibility, fairness, and other human dimensions, and how? What role does listening play?
- How are professional working relationships established, and what are the critical phases of these *partnership building* experiences?
- How do management tools influence actual fishing practices, *compliance*, and thus the resource? What behavioral incentives are regulatory interventions creating?
- What *socioeconomic impacts* have fisheries regulations, cooperative research, and management had on fishermen, the fishing industry, coastal communities, and the broader states and regions?

Appendices

Appendix A: Case Examples, Key Participants, and Descriptions

Cooperative Research

1. *Gear/conservation engineering*

- **Southeast Cooperative Research Program**
  Gerry Scott, director, Sustainable Fisheries Division, Southeast Fisheries Science Center
  Cooperation: Among the National Marine Fisheries Service (NMFS), state fishery agencies, universities, and fishermen.
  Main goal(s): To provide financial assistance to projects seeking to increase and improve cooperative working relationships. The program provides a means of involving commercial and recreational fishermen in the collection of fundamental fisheries information. Collection efforts support the development and evaluation of management and regulator options.
  Web site: http://sero.nmfs.noaa.gov/grants/crp.htm
- **Manomet Center for Conservation Sciences: Composite Net Study, Georges Bank**
  Chris Glass, director of marine fisheries, Manomet Center for Conservation Sciences
  Cooperation: Between Manomet and fishermen.
  Main goal(s): To demonstrate that a new cod end design would be more effective and conservative than what was currently used in mixed-species assessments.
  Web site: http://www.manomet.org/programs/marine/
- **Raised Footrope Trawl Project, Gulf of Maine**
  Mike Pol, senior marine fisheries biologist, Conservation Engineering and Fisheries Dependent Investigations project leader, Division of Marine Fisheries
  Cooperation: Among fishermen, the NMFS, and the Northeast Fisheries Management Council.
  Main goal(s): To reopen the small-mesh whiting fishery by establishing an exempted federal fishery by achieving a reduction in bycatch of regulated species. The development of a raised footrope greatly reduced bycatch but did not meet NMFS bycatch levels to allow exemption.
  Web sites: http://www.mass.gov/dfwele/dmf/publications/rft_whiting_tr3.pdf; http:/www.mass.gov/dfwele/dmf/programsandprojects/conseng/sweepless_report.pdf

2. *Survey research*

- **Northwest Fisheries Science Center Cooperative Research Program**
  Elizabeth Clarke, division director, Northwest Fisheries Science Center
  Cooperation: Among the Pacific States Marine Fisheries Commission, the Oregon Sea Grant, Oregon State University, the acific Whiting Conservation Cooperative, and multiple fishers.
  Main goal(s): To provide opportunities for many sectors of the fishing industry to work cooperatively in many facets of fisheries research.
  Web site: http://www.nwfsc.noaa.gov/research/divisions/fram/westcoastgroundfish westcoastgroundfish.cfm
- **Cooperative Crab Research /Bering Sea Fisheries Research Foundation**
  Gary Stauffer, Alaska Fisheries Science Center, Resource Assessment and Engineering Division

Cooperation: Among the industry (processing plants, 250 vessels), the NMFS, 13 volunteer board members (foundation), the Alaska Department of Fish and Game, and the city of Dutch Harbor.
Main goal(s): To manage the crab fishery in the Bering Sea.
Web site: www.afsc.noaa.gov/race/default.htm

- **Manomet Center for Conservation Sciences: Yellowtail flounder *Limanda ferruginea*, Gulf of Maine**
  Chris Glass, director of marine fisheries, Manomet Center for Conservation Sciences
  Cooperation: Between Manomet and fishermen; among associated fisheries of Maine (24 vessels), the NMFS, the offshore Lobstermen Association, and the coastguard.
  Main goal(s): Fishermen wanted to demonstrate that at certain times of the year, they could target yellowtail flounder in a closed area in Georges Bank with little bycatch.
  Web site: www.manomet.org/programs/marine/

- **Commercial Vessel Acoustic Survey of Coastal Herring Spawning Units**
  Kevin Scheirer, project manager, fisheries biologist, Gulf of Maine Aquarium
  Cooperation: Among the fishing industry, herring processors, the Department of Marine Resources, and the NMFS/Northeast Fisheries Science Center (NEFSC).
  Main goal(s): To conduct a spawning stock survey using commercial fishing vessels equipped with scientific-grade echo sounders, analyze the acoustic data, and make survey results available to the stock assessment and management processes and industry; develop and execute a reliable annual inshore hydroacoustic survey, which complements the NMFS and Fisheries and Oceans Canada (DFO) surveys, which is a critical step in understanding the population dynamics of the coastal Maine herring unit.
  Web site: www.gma.org/herring/

- **SMAST Environmental Assessment: a before-after–control-impact study of the sea scallop aggregations of Georges Bank**
  Kevin Stokesbury, assistant professor, School of Marine Sciences and Technology, University of Massachusetts Dartmouth
  Cooperation: Among the NMFS, the Massachusetts Division of Marine Fisheries, the industry (135 fishing vessels, 74 commercial fishermen), shoreside support businesses, the city of New Bedford, and university academics.
  Main goal(s): Mapping the seafloor using new video technology (video-spatial distribution), stock assessment with new technology, gear design, environmental assessment, life history parameters, and simulation model.
  Web site: www.smast.umassd.edu/Fisheries/Scallops/

- **Cooperative survey of Atlantic surfclams *Spisula solidissima***
  Jim Weinberg, research scientist, NMFS, Woods Hole Lab
  Cooperation: Among commercial fishermen (surfclam industry), Rutgers University, Woods Hole, and the NMFS/NEFSC.
  Main goal(s): To survey the closed area region with the goal to mimic methods used by government survey and look at abundance and biomass of surf clams.
  Web site: nefsc.noaa.gov/nefsc/publications/crd/crd0501/

3. Tagging studies

- **Yellowtail Flounder Tagging Program in Georges Bank**
  Steve Cadrin, research fisheries biologist, NOAA Fisheries
  Cooperation: Among commercial fishermen, shore-side support businesses, state agencies, the NMFS, the Canada DFO, and universities.

Main goal(s): To foster a working relationship between fishermen and scientists and work toward several research goals to be accomplished via a mark–recapture study for yellowtail flounder.
Web site: www.cooperative-tagging.org/

- **Northeast Regional Cod Tagging Program**
  Shelly Tallack, program manager/researcher, Gulf of Maine Research Institute
  Cooperation: Among ~250 fishermen (recreational and commercial), ~25 scientists from Nova Scotia to Cape Cod, state and federal governments, universities, and nonprofits.
  Main goal(s): (1) Improve understanding of cod movements in the Gulf of Maine, Georges Bank, and neighboring New England and Canadian waters; (2) tag 100,000 cod throughout the region using a standard tagging technique; (3) obtain contemporary growth information; (4) encourage the participation of commercial and recreational stakeholders; (5) build an online database with accessibility through a geographic information systems mapping interface; and (6) identify and develop testable hypotheses for further research.
  Web site: www.codresearch.org

- **Cod tagging, University of New Hampshire**
  David Goethel, commercial fisherman
  Cooperation: Between University of New Hampshire and commercial fishermen.
  Main goal(s): To tag cod and look at movement and spatial distribution.
  Web site: www.northeastconsortium.org/articles/cod_tagging.shtml

- **Lake Michigan Yellow Perch Task Group (YPTG)**
  1. *Paul Allen, Ball State University, fisheries biologist, current chair of YPTG*
  2. *John Dettmers, director, Illinois Natural History Survey Lake Michigan Biological Station*
  Cooperation: Among state government departments of natural resources (Indiana, Wisconsin, Michigan, and Illinois), the intertribal regulatory body for fishing tribes (the Chippewa Ottawa Resource Authority), universities, and nonprofits (Great Lakes Fish ery Commission)
  Main goal(s): To develop a multiagency research initiative to identify the likely causes for yellow perch *Perca flavescens* recruitment failure, prioritize research goals toward maintaining self-sustaining stocks of yellow perch.
  Web site: www.glfc.org (http://www.glfc.org/lakecom/lec/LEC_YPTG.htm)

4. Fisheries biology, ecosystem, habitat, oceanography

- **Pacific Northwest Aquatic Monitoring Program**
  Jen Bayer, biologist, facilitator/project coordinator, U.S. Geological Survey (USGS)
  Cooperation: Among tribal entities, state agencies (Washington, Oregon, and California), federal government agencies, and regional entities.
  Main goal(s): To provide a forum for coordinating state, federal, and tribal aquatic habitat and salmonid monitoring programs; improve communication; share resources and data; coordinate compatible monitoring efforts; provide increased scientific credibility and cost effective use of limited funds; and create greater accountability to stakeholders.
  Web site: http://www.pnamp.org/web/content.cfm?WorkGroupID=1

- **Gulf of Maine Closed Area Study, New Hampshire**
  Ray Grizzle, research associate professor, University of New Hampshire (UNH), Marine Program
  Cooperation: Among Northeast Consortium, UNH, National Science Foundation, National Oceanic and Atmospheric Administration (NOAA), Portsmith Coop, and several commercial fishermen

Main goal(s): To assess and monitor the ecological, social and economic impacts from the Gulf of Maine closed area.
Web site: cinemar.unh.edu/fisheries/jeffreys_ledge/index.html

- **Environmental Monitoring of Lobster Traps (eMOLT)**
  1. *Jim Manning, oceanographer, NMFS/NEFSC*
  2. *Patrice McCarron, executive director, Gulf of Maine Lobster Foundation and Maine Lobstermen's Association, Inc.*
  Cooperation: Among NOAA/NEFSC, four lobstermen associations, and a community college.
  Main goal(s): The eMOLT project is a nonprofit collaboration of industry, science, and academics devoted to monitoring the physical environment of the Gulf of Maine and the southern New England shelf.
  Web sites: www.emolt.org and www.gomlf.org/

- **Belle Isle Sturgeon Reef Project, Detroit River**
  1. *Bruce Manny, research fishery biologist, USGS*
  2. *Doug Denison, director of environmental studies, JJR Inc.*
  Cooperation: Among Michigan State University, SmithGroup JJR Inc., DTE Energy Inc., the Michigan Department of Environmental Quality, the Michigan Department of Natural Resources-Fisheries Division, the City of Detroit-Recreation Department, the U.S. Fish and Wildlife Service, the USGS, the Detroit Zoological Institute, the Detroit River Remedial Action Plan Team, the Friends of the Detroit River; the Friends of Belle Isle, the U.S. Army Corps of Engineers, the Greater Detroit American Heritage River Initiative, Sturgeon for Tomorrow (Black Lake, Michigan Chapter); the Essex Region Conservation Authority; the Ontario Ministry of Natural Resources, the Wisconsin Department of Natural Resources, and the Wyandot of Anderdon Nation
  Main goal(s): To enhance a healthy and self-sustaining population of lake sturgeon in the Detroit River by constructing viable spawning habitat and through education, outreach, and monitoring
  Web site: www.miseagrant.umich.edu/sturgeon/background.html

- **Columbia River Basin Regional Research Plan**
  Steve Waste, coordinator, program manager, Northwest Power and Conservation Council
  Cooperation: Among four states (Washington, Oregon, Idaho, Missouri), six federal agencies, and two tribal entities.
  Main goal(s): To cooperatively coordinate and develop regional research priorities for the Columbia River basin and establish a long-term commitment from each group involved.

- **Fishery Resource Grant Program, Effect of Commercial Fishing Pressure on Habitat, North Carolina**
  Alex Ng, commercial fisherman, North Carolina
  Cooperation: Among universities, the NOAA/Sea Grant, the NMFS, and fishermen.
  Main goal(s): To evaluate the effects of repetitive commercial fishing pressures on specific sites; comparison of historical and present day landings of reef fish species in Onslow Bay, North Carolina
  Web site: www.ncseagrant.org/index.cfm?fuseaction=page&filename=FRG_main.html

- **Fishery Resource Grant Program, North Carolina Blue Crab Study**
  Jimmy Nobles, commercial fisherman, North Carolina
  Cooperation: Among commercial fishermen, university/academics, and the NOAA/Sea Grant.
  Main goal(s): The purpose of this project was to determine the impact of North Carolina's Recreational Commercial Gear License (RCGL) holders on the harvest of blue crabs

Callinectes sapidus. Specific objectives were to estimate the total number of crabs harvested by RCGL holders, as well as the amount of effort they put into crabbing and the types of crabbing equipment they used
Web site: www.ncseagrant.org/index.cfm?fuseaction=page&filename=FRG_main.html

- **Fishery Resource Grant Program, North Carolina Shrimp Bycatch Study**
  Denny McCuiston, commercial shrimp fisherman, North Carolina
  Cooperation: Among commercial fishermen, university/academics, and the NOAA/Sea Grant.
  Main goal(s): To assess bycatch generated in the shrimp fishery in the inland waters of North Carolina.
  Web site: www.ncseagrant.org/index.cfm?fuseaction=page&filename=FRG_main.html

- **Port Orford Ocean Resources Team (POORT), Oregon**
  1. *Carla Sowell, natural resource specialist II, south coast port biologist, Oregon Department of Fish and Wildlife*
  2. *Leesa Cobb, communication coordinator and port sampling supervisor, Port Liaison Project*
  3. *Anonymous rockfish fisherman*
  Cooperation: Among Oregon Department of Fish and Wildlife/POORT, fishermen, shoreside support businesses, the NMFS, the Pacific Marine Conservation Council, the Oregon State University Extension, and several nonprofits.
  Main goal(s): To assess the age structure, maturity, and conduct genetic sampling of three fish species from commercial fishing vessels at port.
- **Fisheries Resource Monitoring Program (strategic planning process), Alaska**
  Peggy Merritt, decision analyst, facilitator, resource decision support
  Cooperation: Among fishermen (recreational and commercial), state government, federal government agencies, regional councils, universities, and nonprofit and private organizations.
  Main goal(s): To ensure that the Fisheries Resource Monitoring Program, the Office of Subsistence Management, and the U.S. Fish and Wildlife Service focus on the highest priority information needs for management of federal subsistence fisheries for the next 3–5 years. The plan will guide the call for proposals regarding the following goals: (1) obtain, develop, and improve information to sustain fish populations necessary to provide for subsistence uses; (2) assess and monitor subsistence fisheries; and (3) develop and evaluate effective regulatory and management strategies.

5. Social Science

- **Canadian Maritime Participatory Social Science Fisheries Research**
  Melanie Wiber, professor of anthropology; University of New Brunswick
  Cooperation: Among university social scientists, fishermen associations (two native and three nonnative), and nonprofit community organizations.
  Main goal(s): To focus on the practical needs for social science research within communities involved in resource management and facilitate critical analysis; there are three main areas: (1) new modes of research collaboration, (2) new ways of producing and structuring knowledge, and (3) knowledge transfer and integration at the national and international levels.

- **Bay of Fundy Atlas of Groundfish Spawning**
  Maria Recchia, Bay of Fundy program coordinator for the Extension Department, Center for Community Based Management
  Cooperation: Among commercial fishermen, 14 fishermen's organizations (Bay of Fundy Fisheries Council), the university/academic community, and several environmental organizations.

Main goal(s): To survey fishermen to collect data for mapping groundfish spawning grounds in the Bay of Fundy.
Web site: http://www.gulfofmaine.org/times/winter2002/scientists.htm (for more infor mation)

- **Gulf of Maine Spawning Ground Project, Eastern Gulf of Maine**
  Ted Ames, executive director of Maine Gill-Netters Association (MEGA)
  Cooperation: Between MEGA and Maine Fishermen's Coop.
  Main goal(s): To lobby for more cod research to study restoration strategies in hopes of rebuilding coastal cod stocks.

Cooperative Management

1. Local level

- **Community-Based Fisheries Clam Management Program, Maine**
  Sherm Hoyt, University of Maine, Sea Grant Extension, fisheries outreach coordinator
  Cooperation: Among local clam fishermen in Maine; they have organized themselves and have established a community-based fisheries management program.
  Main goal(s): To manage a sustainable clam resource.
  Web site: www.seagrant.umaine.edu/contact/bio/biosherm.htm
- **Downeast Initiative, Area Management of Coastal Groundfishery, Downeast Maine**
  Ted Hoskins, fisherman, fishing community activist, community leader
  Cooperation: Among fishermen, shoreside support businesses, local and state governments, several fishing councils, universities/academics, nonprofit organizations
  Main goal(s): To develop a pilot project for the area management of the coastal groundfishery, learn from history (habitat degradation, overfishing, pollution, human impacts) and change the way fisheries and the ocean resources are managed.
- **Bay of Fundy Fixed Gear Council**
  Hubert Saulnier, co-chair of The Fundy Fixed Gear Council
  Cooperation: Between the Bay of Fundy Inshore Fishermen's Association and the Maritime Fishermen's Association.
  Main goal(s): To manage so as to ensure the sustainability of our fisheries in the Bay of Fundy ecosystem as a whole and, also, to be able to acquire more allocations and obtain more authority over decision making.

2. Regional—multistate/provincial

- **Mississippi Interstate Cooperative Resource Association**
  Jerry Rasmussen, coordinator and professional fishery biologist, USFWS
  Cooperation: Among 28 state governments bordering in the Mississippi River basin, four federal agencies, and two tribal entities.
  Main goal(s): To improve the conservation, development, management and utilization of interjurisdictional fishery resources in the Mississippi River basin through improved coordination and communication among the responsible management entities.
  Web site: http://wwwaux.cerc.cr.usgs.gov/MICRA/
- **Northwest Power and Conservation Council Subbasin Planning**
  Lynn Palensky, subbasin planning coordinator
  Cooperation: Among the Bonneville Power Administration, the USFWS, the U.S. Forest

Service, the NOAA, state government agencies, 13 tribal entities, watershed councils, re covery boards, counties, cities, conservation districts, and landowners.
Main goal(s): To protect, mitigate and enhance the fish and wildlife of the Columbia River basin affected by hydropower dams while also assuring the region an adequate, efficient, economical, and reliable power supply.
Web site: www.subbasins.org

- **Atlantic States Marine Fisheries Commission (ASFMC)**
  Vince O'Shea, executive director, Atlantic States Marine Fisheries Commission
  Cooperation: Among members of the ASFMC are the 15 Atlantic Coastal States from Maine to Florida, including Pennsylvania. Partners include the NMFS and the USFWS. Also a cooperation/partnership with the three regional fishery management councils, and stakeholders.
  Main goal(s): To working towards healthy, self-sustaining populations for all Atlantic coast fish species or have successful restoration programs well in progress by 2015.
  Web site: www.asmfc.org

- **Regional Fishery Management Councils (8)**
  Emily Menashes, director, Office of Sustainable Fisheries, NMFS
  Cooperation: Members represent stakeholders for each state within the geographic area served by a council.
  Main goal(s): To manage fishing and fisheries in the U.S. exclusive economic zone, also to write fishery management plans.
  Web site: http://www.nmfs.noaa.gov/councils.htm

3. *International examples, multigovernmental*

- **Great Lakes Fishery Commission (GLFC) – Joint Strategic Plan for Managing Great Lakes Fisheries**
  Marc Gaden, GLFC, communications officer and legislative liaison, facilitator
  Cooperation: Among tribes, eight states, provinces around the Great Lakes, federal governments (the USGS, NOAA Fisheries, the USFWS, and Fisheries and Oceans Canada), several consulted and involved committee of advisors, and some fishermen.
  Main goal(s): To maintain a consensus-based plan for research needs and goals, to perform joint assessments and set common quotas, and to make sure the plan is followed and implemented (consultation, hold forums, accountability, information sharing, and data analysis).
  Web site: www.glfc.org

- **International Pacific Halibut Commission (IPHC)**
  Bruce Leaman, executive director, IPHC
  Cooperation: Among Canadian and U.S. governments, fishermen, shoreside support businesses, and two regional fisheries management councils.
  Main goal(s): To conduct single species stock assessments, several scientific experiments focused on basic halibut biology, fishing surveys, and field sampling at major ports.
  Web site: www.iphc.washington.edu

- **Shetland Islands, UK—Rights-Based Management Example**
  John Goodlad, fisheries consultant and fish farmer, past chief executive, Shetland Fish Producers Organisation Ltd.
  Cooperation: Between fishermen's organizations and a local development agency.
  Main goal(s): To give fishermen the opportunity to manage the fisheries based on a quota system; then, member vessels can share quota and manage in a manner best suited to local market conditions.
  Web site: www.users.zetnet.co.uk/sfa/

- **New Zealand Rock Lobster Management**
  1. *John Annala, former New Zealand Ministry of Fisheries, stock assessment manager*
  2. *Daryl Sykes, executive officer, former fisherman, NZ Rock Lobster Industry Council, Ltd.*

  Cooperation: Among rock lobster fishermen, quota holders, processors, exporters, research consultants, and the New Zealand Ministry of Fisheries.
  Main goal(s): To coordinate the research and management needs of the rock lobster fishery in nine regions through local commercial stakeholder organizations.
  Web sites: http://www.seafood.co.nz/industry/otoz/rlicoun/default.asp; http:/www.seafood.co.nz/industry/aton/cra2rock.asp

Appendix B: Interview and Survey Questions

Case Study Questions for the Symposium *Partnerships for a Common Purpose: Cooperative Fisheries Research and Management*

Note: Please share as much information as you like, space is NOT limited to the box sizes. Fill in the information within this document and return to Alesia Read at aread@fisheries.org. The information you share will be held confidential and will not be associated with your name. The case studies will be used to prepare symposium participants as an online resource, as well as included in the final symposium proceedings.

Goal: To gather information about individual cases of cooperative research or cooperative management to understand

- what makes cooperative research and management projects successful, and
- critical lessons learned and implementation tips.

Please provide a brief paragraph description of the project, its goals and/or mission statement.

Does the project have a Web site? If not, is there a report or written overview that you could provide a copy of?

Scale:

- ☐ Local
- ☐ Multiple local communities
- ☐ State-wide
- ☐ Multi-state
- ☐ National
- ☐ Other: ______________

Who are the partners? Are they mostly paid staff or volunteers? Please list the organizations represented. Who are the leaders?

Who do the partners represent? Please check all that apply

- ☐ Commercial fishermen—type of fishery:
- ☐ Recreational fishermen/charter and party boats
- ☐ Shore-side support businesses
- ☐ Local government—list departments/programs:
- ☐ State government—list agencies/departments:
- ☐ National Marine Fisheries Service/federal government—list agencies:
- ☐ Fisheries management council
- ☐ University/academic community
- ☐ Environmental organizations
- ☐ Nonprofit, community organizations—types:
- ☐ Other

What is your own role in the effort? How much time (%) do you spend on the project per month?

How is the project funded?

When/how did the project start? Is it ongoing or completed?

Why did you, personally, decide to participate in this project? What are your own interests and incentives for participating?

How often do you consult, coordinate, or work with these partners (e.g., daily, weekly, monthly, quarterly, annually)?

How is communication and coordination maintained *among the partners*? Please check all that apply and estimate frequency.

- ☐ E-mail/list serve. Frequency: ___________
- ☐ Newsletters/mail. Frequency: ___________
- ☐ Phone/conference call. Frequency: ______
- ☐ Opportunistic face-to-face. Frequency: ___
- ☐ Regular meetings. Frequency: _______
- ☐ Web page. Update frequency: _______
- ☐ Reports/publications. Frequency: ____
- ☐ Dock/boat ramp. Frequency: _______

How does the project communicate with *external audiences*? Please check all that apply and estimate frequency.

- ☐ E-mail/list serve. Frequency: ___________
- ☐ Newsletters/mail. Frequency: ___________
- ☐ Phone/conference call. Frequency: ______
- ☐ Commercials. Frequency: ____________
- ☐ Regular meetings. Frequency: _______
- ☐ Web page. Update frequency: _______
- ☐ Reports/publications. Frequency: ____
- ☐ Newspaper articles. Frequency: ______

What communication mechanisms do you personally find most effective for *internal* communication among the partners? Why?
External communication? Why?

Do you consider the project successful? Yes ☐ No ☐
In what ways has the project been successful?
In what ways has it not been successful or could be improved?
How has the project sustained itself and kept up momentum?

What have been the biggest obstacles or challenges? Please check all that apply and indicate its level of significance by highlighting the corresponding level:

	Level of significance				
	Somewhat		Moderate		Very
☐ Coordinating the partners.	1	2	3	4	5
☐ Obtaining the needed information.	1	2	3	4	5
☐ Managing the information.	1	2	3	4	5
☐ Understanding all the complex information.	1	2	3	4	5

□ Overcoming divergent interests of partners.	1	2	3	4	5
□ Obtaining sufficient financial resources.	1	2	3	4	5
□ Ineffective communication.	1	2	3	4	5
□ Mistrust among partners.	1	2	3	4	5
□ Finding feasible solutions that meet partner needs.	1	2	3	4	5
□ Handling hardball tactics, destructive behavior.	1	2	3	4	5
□ Dealing with external/outside opponents.	1	2	3	4	5
□ Implementing our decisions internally.	1	2	3	4	5
□ Other?	1	2	3	4	5

Please describe:

How has the project overcome these obstacles?

What has been the biggest challenge for you personally participating in this project? Please check all that apply and indicate its level of significance by highlighting the corresponding level:

	Level of significance				
	Somewhat		Moderate		Very
□ Remaining committed and motivated.	1	2	3	4	5
□ Managing all the information.	1	2	3	4	5
□ Learning all that I need to make informed decisions.	1	2	3	4	5
□ Dealing with difficult people.	1	2	3	4	5
□ Communicating with other partners.	1	2	3	4	5
□ Finding the time to participate.	1	2	3	4	5
□ Other?	1	2	3	4	5

Please describe:

How have you overcome this challenge?

What have you personally found most rewarding about participating in this project?

LESSONS LEARNED: Based upon your experience with the project, many would benefit from hearing what you have learned and what advice you would give other people who are considering a similar project? Particularly regarding

Why do you think people should/should not collaborate?

What is particularly important at critical stages of the process?

Getting started?

Organizing and managing?

Making decisions?

Assessing progress?

In hindsight, what would you have done differently if you had the whole project to do over?

Other lessons that you think are important?

LAST: What else do you think we need to know about this project that we have not asked?

Who else involved with this project should we ask to fill out this survey? Please provide a name, phone number, and e-mail address for project participants.

Finally, please tell us a little about yourself:

Position and occupation: ______________________________

Education level: ______________________________

Paid staff or volunteer: ______________________________

Length of time with your organization: ______________________________

Gender: Female □ Male □ Year of birth: __________

Thank you for taking the time to fill out this questionnaire. The information you share will benefit many participants in the *Partnerships for a Common Purpose: Cooperative Fisheries Research and Management* symposium at the 2005 American Fisheries Society's annual meeting. In addition, the proceedings will include the case studies and will be a valuable resource for others considering cooperative research or cooperative management initiatives. We hope that you will attend the 2005 American Fisheries Society meeting and participate further in the symposium. If you have any further questions or concerns about this project, please contact, Alesia Read, AFS/Sea Grant Fellow, at 301-897-8616, ext. 222, or over email at aread@fisheries.org. After completing the questionnaire, please save the file and return it to Alesia.